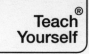

Understand Physics

Jim Breithaupt

For UK order enquiries: please contact Bookpoint Ltd,
130 Milton Park, Abingdon, Oxon OX14 4SB.
Telephone: +44 (0) 1235 827720. Fax: +44 (0) 1235 400454.
Lines are open 09.00–17.00, Monday to Saturday, with a 24-hour
message answering service. Details about our titles and how to
order are available at www.teachyourself.com

For USA order enquiries: please contact McGraw-Hill Customer
Services, PO Box 545, Blacklick, OH 43004-0545, USA.
Telephone: 1-800-722-4726. Fax: 1-614-755-5645.

For Canada order enquiries: please contact McGraw-Hill Ryerson
Ltd, 300 Water St, Whitby, Ontario L1N 9B6, Canada.
Telephone: 905 430 5000. Fax: 905 430 5020.

Long renowned as the authoritative source for self-guided learning –
with more than 50 million copies sold worldwide – the **Teach Yourself**
series includes over 500 titles in the fields of languages, crafts, hobbies,
business, computing and education.

British Library Cataloguing in Publication Data: a catalogue record
for this title is available from the British Library.

Library of Congress Catalog Card Number: on file.

First published in UK 2002 by Hodder Education, 338 Euston Road,
London NW1 3BH.

First published in US 2002 by The McGraw-Hill Companies Inc.

This edition published 2010.

Previously published as *Teach Yourself Physics*

The **Teach Yourself** name is a registered trade mark of
Hodder Headline.

Typeset by MPS Limited, A Macmillan Company.

Printed in Great Britain for Hodder Education, an Hachette UK
company, 338 Euston Road, London NW1 3BH, by
CPI Group (UK) Ltd, Croydon, CR0 4YY.

Hachette UK's policy is to use papers that are natural, renewable
and recyclable products and made from wood grown in sustainable
forests. The logging and manufacturing processes are expected to
conform to the environmental regulations of the country of origin.

Impression number   10 9 8 7
Year   2014 2013

Acknowledgements

I would like to thank my family for their support in the preparation of the book, particularly my wife, Marie, for secretarial support and cheerful encouragement. I am also grateful to the publishing team at Hodder Education, in particular Helen Hart who initiated the project, Catherine Coe who edited and coordinated the original production of the book and Harry Scoble and Helen Rogers who worked on this new edition.

Contents

Meet the author

Welcome to *Understand Physics*!

I am a physics writer with extensive experience of teaching physics in schools and colleges. In addition to *Understand Physics*, I have written *Einstein – A Beginner's Guide* and a considerable number of highly-regarded school and college text books including *New Understanding Physics for Advanced Level* and the AS level and A2 physics textbooks for *AQA Physics Specification A* as well as *Physics for IGCSE* , *Key Science Physics for GCSE*, *AQA GCSE Physics* and the physics content of *AQA GCSE Science* and *AQA GCSE Additional Science*. I have also written *Foundations in Physics* for university foundation courses in physics and physics-related subjects. My physics text books are used in many countries and *Einstein – A Beginner's Guide* has been translated into several languages including Spanish and Mandarin. I also wrote the *A-Z guide to physics* for the *Advancing Physics* course sponsored by the U.K. Institute of Physics. I am an experienced A level physics examiner and frequently contribute to in-service courses for physics teachers.

Physics is the subject we all rely on for every step we take and every breath we breathe. This book is intended to introduce you to the essential ideas and rules of physics, presented to make the subject as easy as possible. We will be looking at ideas and experiments that stretched the minds of scientists and philosophers in the past, but they were pioneers and we are not, and so we do not need to explore all the dead ends along the way to the frontiers of the subject. At times, we will pause on our journey to take in important ideas and facts that are needed to build up an overview of the subject and make further progress towards the frontiers. We will also look at the methods and skills used in scientific investigations and how ideas were developed from investigations and then tested and used. So that's a few clues about how we will be making the journey that lies ahead. Now let's see what the route will cover.

Jim Breithaupt, 2010

Only got a minute?

Physicists tend to be a bit unconventional because physics is a very varied subject and physicists like to be able to turn their knowledge and skills to almost any situation. Richard Feynman (1918–88) was one of the most creative physicists of the twentieth century. Following wartime duties on the Manhattan Project, which created the atom bomb, he made his name through developing a new understanding of the force between charged objects. After the 1986 *Challenger* space shuttle disaster Feynman gave crash investigators a simple yet historic demonstration showing what had caused the disaster: an O-shaped sealing ring on the fuel line had cracked when cooled to the very low temperature associated with the type of fuel used. So physicists get to grips with how and why objects and materials behave as they do and why objects interact and move as they do. In short, physics is about matter, which is what material objects are made from, and it is about energy, which is to do with how objects interact and move.

However, to reduce physics to the study of matter and energy is to undervalue the subject, as physicists can be found applying their skills and knowledge in a vast range of situations beyond the physics laboratory, for example, in archaeology, communications, computing, engineering, hospitals, business and finance, sport and leisure, in fact just about everywhere. The World Wide Web (WWW), which has revolutionized business and commerce, was set up by physicists to enable them to communicate around the world and discuss scientific papers electronically. The latest generation of body scanners were developed from discoveries by physicists working on a phenomenon known as magnetic resonance. These scanners have revolutionized the study of the brain, making it possible for scientists to see how different parts of the brain respond to different experiences and situations.

5 Only got five minutes?

The impact of physics

We live in the Scientific Age, which has brought immense benefits to us all. Compare your lifestyle with that of someone two centuries ago and you will realize that the vast majority of people then lived in abject poverty, dependent for their very lives on the vagaries of the weather, which could ruin harvests and hit transportation of food and other essential commodities. The Scientific Age has brought power on demand, so we need not rely on wood or coal for cooking and heating our homes. Travel is much quicker, cheaper and far more convenient than two centuries ago when sea journeys took weeks and overland journeys days. People are healthier and fitter and live longer now than then, thanks to improvements such as clean water, better nutrition and medical drugs. All these improvements have come about because of discoveries made by scientists. To appreciate the impact of physics, consider just two discoveries made in physics in the nineteenth century which revolutionized life in the twentieth century.

Michael Faraday (1791–1867) discovered how to generate electricity and thereby laid the foundations for the electrical supply industry and for electrical communications. Faraday gained an appetite for science in his first job as a bookbinder's apprentice. As well as binding the books, he read them avidly and learned about electricity from them. He wrote to Sir Humphry Davy at the Royal Institution in London and gained a position as an assistant as a result. His research into electromagnetism led to his famous discovery of how to generate electricity using magnets. He developed great insight into the link between electricity and magnetism and discovered the fundamental principles that led to the invention of the alternating current generator, the transformer, the electric motor and the electric telegraph. The electricity supply industry developed in the latter half of the nineteenth century from Faraday's discoveries. By the first decades of the twentieth century, most homes in Britain and other industrial nations had been wired up to a local electricity supplier and nationwide grid systems for electricity distribution were established. Nowadays we take our electricity supplies for granted, unless a power cut occurs and we are thrown temporarily back into the pre-electricity era.

When Faraday demonstrated his discoveries to an invited audience at the Royal Institution, he was asked, 'What use is electricity, Mr Faraday?' He replied with another question, 'What use is a new baby?' No one can tell what can grow from a new discovery.

James Clerk Maxwell (1831–79) was a mathematical physicist who used Faraday's ideas to devise a theory that light is an electromagnetic wave. He showed that an electric wave and a magnetic wave can travel together through space at a speed of 300, 000 kilometres per second, the same as the speed of light. He therefore deduced that light must be an electromagnetic wave. He already knew that the spectrum of light covers a continuous range of colours; namely red, orange, yellow, green, blue and violet. His theory showed that all the colours of light are electromagnetic waves of different wavelengths. When he put forward this theory in 1865, he knew that infrared radiation lies just beyond the red part of the visible spectrum and that ultraviolet radiation lies just beyond the violet part of the spectrum.

His theory of electromagnetic waves thus accounted for infrared and ultraviolet radiation as well as light. However, he predicted the existence of electromagnetic radiation outside the known spectrum of infrared radiation, light and ultraviolet radiation. Two decades later, Heinrich Hertz discovered radio waves and he was able to show by measurement that they travel at the same speed as light, thus providing clear evidence that radio waves are electromagnetic waves. Hertz's discoveries were taken up by other scientists including Marconi, who led the way forward into the use of radio waves in wireless communications. The enormous range of the radio spectrum is now used for many purposes, including radio and TV broadcasting, mobile phone communications, radar systems, satellite communications and navigation. Maxwell's theory of electromagnetic waves provides the theoretical basis for all these forms of communications.

Many more examples can be found of the way in which fundamental discoveries in physics have dramatically altered the way we live. The discovery of the atomic nucleus by Ernest Rutherford and the invention of the transistor are two examples of experimental discoveries in physics that have led to the growth of new industries. Albert Einstein's theories of relativity and the development of quantum mechanics by Niels Bohr and others are examples of theories that overturned established ideas of physics and provided radical new approaches in the subject. We will meet these and other shattering new discoveries and ideas later on in this book. However, do not imagine that such dramatic changes are all history now. Physicists continue to produce astonishing new discoveries and amazing new ideas. Antimatter, black holes and superstrings are but a few of the latest hot topics in physics. The explosive growth of the World Wide Web is another result of physicists at work barely a decade ago, in this case developing an electronic method of keeping in touch with each other which was then taken up at an astounding pace by people and businesses in all walks of life.

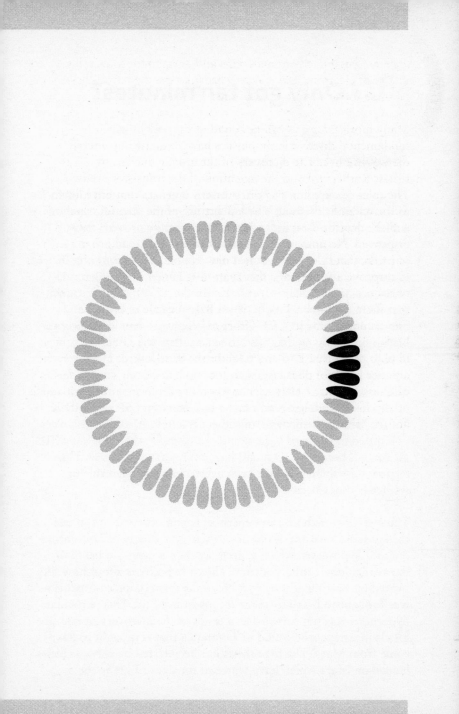

10 Only got ten minutes?

The golden rule of science

The methods used by physicists follow a general pattern known as the scientific method. The key feature of the scientific method is that scientific theories and laws can never be proved, only disproved. No amount of experimental evidence can prove a scientific theory or law; yet just one reliable experiment is sufficient to disprove any scientific theory or law. Forget this golden rule of science and disaster is inevitable. In the 1990s, concerns arose that there could be a link between BSE in cattle in Britain and human deaths due to CJD. Other EU countries banned imports of British beef. Top medical and food scientists were unable at the time to prove that BSE and CJD deaths were linked. However, the absence of proof does not mean that no link exists. The golden rule was forgotten, that science works by disproving theories and predictions. Finding no proof of a link between BSE and CJD is not the same as disproving that there is a link. Also, the discovery of a different cause of CJD would not disprove that BSE and CJD are linked, because CJD might have more than one cause. This terrible tragedy might have been averted if the golden rule of science had not been forgotten.

The scientific method uses experiments and observations to test the laws and theories of physics. A scientific theory or law starts out as a hypothesis, which is no more than a proposal based on the available scientific evidence. Thus a hypothesis about how life started on Earth might be that life in the form of micro-organisms was brought to Earth by some means from Mars. This particular hypothesis was put forward as a result of the discovery of microbe fossils in a meteorite found in Antarctica that is thought to have come from Mars. The hypothesis has been tested by robot vehicles landed on Mars which have searched for signs of life in the

Martian soil and atmosphere. No positive findings have yet been reported and so the hypothesis remains just that, although the discovery of water ice in the Martian soil by the Mars Phoenix Lander in July 2008 suggests that life may once have existed on Mars, and further space missions to Mars are being planned. If microbes, past or present, were discovered on Mars, this would boost the status of the hypothesis sufficiently to make it into a theory. The theory that life on Earth arrived from Mars would be strengthened if microbe fossils were found on Mars similar to the ones found in Antarctica. The scientific evidence in favour of the theory would then be very strong, but it would not be possible to say that the theory is true, because life on Earth might have developed before the arrival of the microbe fossil from Mars. No amount of experimentation could ever prove the theory, but successive tests could reinforce it and boost its status.

> Visit the NASA website to learn more about the Mars Phoenix Lander and future plans for the search for life on Mars.

The theory of continental drift developed from a hypothesis put forward by Alfred Wegener in 1915. He realized that the similarities between the eastern coastline of South America and the western coastline of Africa could mean that they were once joined together and then drifted apart. He discovered many similar fossils, plants and animals in the two continents and so he put forward the hypothesis that the continents were once all joined as a single land mass. Wegener was unable to find a convincing source of power to make the continents drift apart, but his ideas were taken up by Arthur Holmes in 1931 and he showed that heat released inside the Earth, due to radioactive decay, could provide the necessary power. Further observations and measurements over several decades showed that the outer layer of the Earth is made up of separate pieces which fit together to make a sphere. These pieces, known as tectonic plates, carry the continents, which are known to drift across the surface as the plates gradually move. Earthquakes and volcanoes occur where the plates are in collision. Wegener's hypothesis thus developed into the theory of plate tectonics, as more and more scientific evidence accumulated in

support of the ideas. Further scientific evidence in support of
the theory would reinforce its status even more. However, the
theory would need to be discarded or drastically altered if reliable
scientific evidence was ever found against it. No theory can ever be
proved, as a single reliable experiment in future could disprove it.

SCIENTIFIC BLUNDERS

The history of science has many cul-de-sacs where discarded
theories lie abandoned and forgotten. Some of these theories were
once held aloft as part of the litany of scientific wisdom.
For example, over two centuries ago, scientists believed that heat is
a fluid that flows from hot objects to cold objects, like water flowing
along a river. The fluid was called 'caloric' and it gave rise to the
calorie as the unit of heat. The supporters of the caloric theory
were undaunted when they were asked to explain where the heat
comes from when you rub your hands together. They imagined that
tiny particles created by the rubbing action released caloric in the
process. The caloric theory was opposed by scientists who reckoned
that heat is a form of energy. The caloric theory was disproved in
1799 by Sir Humphry Davy. He showed that when two blocks of
ice below freezing point were rubbed together, water was produced
as a result of the ice melting. He demonstrated that the melting of
ice occurs because of friction between the two surfaces in contact.
Energy must be supplied to the blocks to make them slide over each
other and this energy causes the ice to melt. No mention was needed
of caloric nor of particles rubbed off the ice.

Cold fusion is a more recent example of a scientific theory that bit
the dust. When the nuclei of two light atoms are fused together,
energy is released because the two nuclei bind together to form a
single nucleus. This process takes place in the core of a star and
is responsible for the energy radiated from the star. The immense
pressure at the centre of a star forces light nuclei together and
makes them fuse. Over the past four decades, scientists have tried
without success to make a fusion reactor which would produce
energy on a large scale at a steady rate. The particles in such a
reactor would have to collide at enormous speed to fuse together,

but at such speeds they are very difficult to control. Two decades ago, a breakthrough was claimed by scientists working on an entirely different approach to the fusion problem. They reckoned they had discovered fusion taking place in the spaces between the surface atoms of a certain metal when small atoms were in contact with the metal under particular conditions. The process was referred to as 'cold fusion', because it seemingly did not require the enormous temperatures necessary for fusion in stars. Other scientists disputed that the release of energy was due to fusion and attempted without success to reproduce the process of cold fusion. They concluded that the energy released was not due to fusion and the claim was rejected. A successful fusion reactor, hot or cold, would be a major step forward in providing energy without pollution and without producing greenhouse gases such as carbon dioxide. Cold fusion proved to be a claim too far.

Strange ideas

Do not be too harsh on scientists whose claims are rejected. In 1905, Albert Einstein produced two scientific papers that revolutionized physics and changed our understanding of nature dramatically. One of these two papers was about the nature of light, which Einstein showed to be in the form of packets of energy he called photons. His other paper, his theory of relativity, proved that absolute space and absolute time do not exist and that mass and energy are linked. We will meet these ideas later. Both papers were revolutionary, yet Einstein was to wait over 15 years before he was awarded the Nobel prize for physics. Why such a long time? The prize is awarded for discoveries that benefit the human race. More than a decade passed before sufficient evidence was available to support Einstein's theories of relativity and light. In recent years, experimental evidence has been discovered for the amazing theories about black holes, developed by theoreticians such as Stephen Hawking. These strange objects are cosmic crushers from which nothing, not even light, can escape. A black hole would be likely to destroy human life as well as our planet.

Maybe there has been no Nobel prize for Hawking yet as, so far, knowledge of black holes seems to have brought few benefits. However, you could argue that the knowledge that helps us to avoid a black hole would definitely be a benefit to the human race! We will return to the subject of black holes later in this book.

Science in general, and physics in particular, is about understanding nature from the smallest scale, deep inside the atom, to the largest scale, stretching to the edge of the Universe. The general method used in science is to look for patterns in observations and experimental data and formulate a hypothesis about the patterns or data. The hypothesis is then used to make a prediction, which can be tested by appropriate experiments. If the results of the experiment confirm the prediction, the hypothesis is used to form a theory, which is tested by more experiments. If the results do not confirm the prediction, the hypothesis must be modified or abandoned. The more experiments that support the theory, the stronger the theory becomes. The Principle of Conservation of Energy is a theory that has withstood many tests. The Principle states that energy is always conserved in any change. The total energy after a change is the same as the total energy before the change. This key principle did not seem to work in a radioactive process known as beta emission. Energy seemed to disappear without trace in this process. It was concluded that either the Principle of Conservation of Energy does not work in this process, or a particle that could not be detected carries away the missing energy. This elusive particle was given a name, the neutrino. Twenty years after the problem was raised, neutrinos were at last detected and the Principle of Conservation of Energy was saved.

Physics is about probing the laws of nature and challenging the status quo. This book is intended to explain the key ideas of physics at the present time, but be in no doubt that new and strange ideas lie ahead. More science has been discovered in the last half century than ever before. Physics is an exciting subject for anyone who thrives on changes and challenges.

1

Starting physics

In this chapter you will learn:

- *why physics has always been a very important subject, from thousands of years ago right to the present time*
- *why Archimedes became a legend when he carried out the first recorded example of non-destructive testing*
- *how to measure the density of an object.*

Physics has always been at the leading edge of human thought, not just in recent times but also before the Scientific Age, which began about four centuries ago. The theories of science that were held to be true long ago might seem very odd when we compare them with our present knowledge. We must not forget though that the scientists of ancient times, or natural philosophers as they were then called, came up with ideas and theories that were astonishingly sophisticated in comparison with other aspects of life in those times. In this chapter we will look at some of the physics ideas from long ago to see just how these ideas were developed and used. Then we will look at some important skills needed in physics, before moving on to use these skills to carry out density tests.

Physics in practice

BEFORE THE SCIENTIFIC AGE

Many theories of science before the Scientific Age were based on the assumptions that the Earth was at the centre of the Universe and that living beings were created by one or more superior beings who designated a special role for humans. The idea that humans evolved from apes over thousands of centuries found little favour, as it omitted the role of a creator. Theories about the natural world were usually chosen on grounds we would consider unscientific, and selected facts were used to support the theories. Other facts that did not match the theories were discarded as unreliable or imperfect. Not surprisingly, alchemy and astrology were two major strands of scientific endeavour before the Scientific Age. For example, attempts to turn lead into gold or to predict events occupied the working lives of many individuals, undoubtedly financed by rich and powerful patrons who imagined that they would amass further wealth as a result of such activities.

Science as a recorded activity flourished in the culture of Ancient Greece and the Mediterranean civilizations which developed from Greece. The idea that matter is composed of '**atoms**' too small to be seen individually was a theory put forward by Democritus (470–400 BCE). Two centuries later, Aristarchus put forward the theory that the Sun was at the centre of the Universe. In the next century, Archimedes (287–212 BCE) made important discoveries in mechanics and mathematics. The importance of the scientific heritage bequeathed by these and other natural philosophers of the ancient Mediterranean civilizations is undisputed. Indeed the impact of one particular natural philosopher, Aristotle (382–322 BCE), was to influence the conduct of science for many centuries, right through to the beginning of the Scientific Age.

Aristotle was trained by the philosopher Plato, who founded the Academy in Athens. Aristotle accepted Plato's theory of ideas as 'eternal patterns' that lie beyond the natural world. However, he recognized the importance of observations in formulating ideas about nature. Where observations could not be used to decide between competing theories, Aristotle rejected any theory that did not support the accepted ideals and overall philosophy developed by Plato. For example, Aristotle rejected Aristarchus' theory that the Sun, not the Earth, was at the centre of the Universe. There was no observational evidence that the Earth was rushing through space round the Sun and no evidence either to support the associated idea that the Earth is spinning. Nor would Aristotle accept the idea that matter is composed of atoms. He rejected the idea of atoms in favour of the theory that all matter is made of the four elements: earth, water, air and fire. This older theory fitted the idea that the Earth is at the centre of the Universe, water in the seas lies above the Earth, air lies above the seas and fire is in the heavens.

Insight

Aristotle recognized the importance of observations in support of his ideas about nature and his overall philosophy.

Aristotle shaped science into a coherent set of ideas that were consistent with the prevailing world-view, namely that the Earth is the centre of the Universe and therefore a special place has been accorded to the human race by the creator of the Universe. However, Aristotle's rejection of theories that did not fit in meant that interesting ideas were not followed up and investigated further. Aristotle stamped his method on science so firmly that it lasted for over 15 centuries. His approach of picking facts to support accepted theories dominated the way science was conducted long after his death, from Ancient Greece, through the Roman Empire, the Dark Ages and into the Middle Ages. Perhaps Aristotle's authority enabled science to survive over this long period, especially through the Dark Ages. The Church promoted Aristotle's scientific method, in particular the model of the Universe developed by Ptolemy a century after Aristotle. Ptolemy held that the Earth was at the centre of the Universe, with the Sun and Moon moving round it in different circular orbits. Each planet moved round in a circle whose centre moved round on its own circular orbit. This model explained the movement of the Sun, Moon and planets and fitted in with the concept of the 'geocentric' universe. We will look later in this chapter at the bitter struggle between Galileo and the Church which eventually led to the overthrow of Ptolemy's model.

Figure 1.1 Ptolemy's model of the Solar System.

ARCHIMEDES, THE FIRST GREAT SCIENTIST

Archimedes was born in Sicily, where he worked under the patronage of King Hiero, the island's ruler. Archimedes kept in touch with the scholars of Alexandria in Egypt, where one of his inventions, the water screw, was put to practical use to raise water from the river Nile. This device consists of a tight-fitting screw in a cylinder. With the cylinder in water at its lower end, turning the screw raises water up the cylinder from the lower end. Archimedes' screw allowed farmers to irrigate the land bordering the river Nile.

Archimedes also discovered the principle of the lever and worked out how a **force** could be increased using a lever. No doubt you have used the lever principle to move an object that is difficult to shift. For example, when you use a bottle opener as a lever to remove the cap from a bottle, the force you apply to the bottle opener causes a much larger force to act on the cap. The applied force on the lever is referred to as the effort. The lever pivots about a fixed point or fulcrum as shown in Figure 1.2. The force of the lever on the cap is much greater than the applied force because it acts much nearer to the fulcrum than the applied force does.

Figure 1.2 (a) Using a lever. (b) Balancing a beam.

Archimedes investigated the equilibrium conditions for a beam balanced at its centre of gravity with a **weight** supported on each side of the beam. He found that the distance of each weight to the fulcrum was in inverse proportion to the amount of weight.

▶ *For two objects of equal weight, the distances from the fulcrum must be equal for the beam to be in equilibrium.*
▶ *For two objects of weights W_1 and W_0, the distances d_1 and d_0 of the weights from the fulcrum are in inverse ratio to the weights. In other words:*

$$\frac{d_0}{d_1} = \frac{W_1}{W_0}$$

If one of the objects is of known weight, by measuring the distances d_1 and d_0 the other weight can be calculated using the equation above. For example, suppose one of the objects is a standard weight at a distance of 0.50 m from the fulcrum; if the beam is balanced by an object of unknown weight at a distance of 0.25 m from the fulcrum, the unknown weight must be exactly twice as heavy as the standard weight.

Insight

You don't need to look very far to find a lever. The bones in your arms and legs are levers that pivot about your joints and are moved by your muscles.

Balancing tests

Place a pencil on a flat surface and balance a ruler on the pencil at right angles to the pencil.

▶ Place a coin on the ruler near one end. Rebalance the ruler by positioning an identical coin on the ruler near the other end. You should find the two coins are at the same distance from the pencil when the ruler is balanced again.
▶ Place a third identical coin on top of one of the coins already on the ruler. Reposition this 'double coin' to balance the ruler again. You should find that the distance from the single coin to the pencil is twice the distance from the double coin to the pencil.

WEIGHTS AND MEASURES

The scientific system of units is known as the **SI system**. This was agreed internationally several decades ago. Before this agreement was reached, different systems of units were used in different countries. In earlier times, before national systems were established, local systems developed by traders varied according to custom and practice.

The five base units of the SI system are:

1 *the* **second**, *which is defined as the time taken for a specified number of vibrations of a certain type of atom in an atomic clock.*

2 the **metre**, *which is defined as the distance travelled by light in a vacuum in a specified time.*

3 the **kilogram**, *which is the unit of mass. This is defined as the quantity of matter in a block of platinum kept in the Bureau of International Weights and Measures (BIPM) in Paris. Standard masses kept in other scientific laboratories are measured by comparison with the standard one kilogram mass in Paris. The weight of an object is the force of gravity on the object. Because the weight of an object is in proportion to its mass, a beam balance as described earlier may be used to compare the masses of any two objects.*

4 the **kelvin**, *which is the scientific unit of temperature. See p. 68.*

5 the **ampere**, *which is the scientific unit of electric current. See p. 103.*

All other scientific units are derived from these five base units. Some examples of derived units are given in the following table. Note that in writing the symbol for a unit, the historical use of the oblique symbol '/' for 'per' (as in metre/second) is usually replaced in more advanced work by a numerical superscript after the symbol (as in ms^{-1}).

Quantity	Definition	Unit
Area of a rectangle	length × breadth	square metres, m^2
Volume of a box	length × breadth × height	cubic metres, m^3
Density	mass per unit volume	kilogram per cubic metre, kg/m^3 or $kg\,m^{-3}$
Speed	distance moved per unit of time	metre per second, m/s or ms^{-1}

Insight

The metre is one thousandth (1/1000) of 1 kilometre, where 1 kilometre was originally defined as one ten-thousandth (1/10 000) of the distance from the Earth's North Pole to the equator. The distance round the Earth is about 40 000 km.

Powers of ten

Scientific measurements and calculations involve values which can range from extremely small to enormously large. For example, the diameter of an atom is about 0.0000000003 metres and the distance from the Earth to the Sun is about 150 000 000 000 metres. Such values are usually written in **standard form** as a number between 1 and 10 multiplied by an appropriate power of ten. For example, the distance from the Earth to the Sun written in this way is 1.5×10^{11} m, where m is the abbreviation for metres and $10^{11} = 100\ 000\ 000\ 000$. Note that 10^{11} is quoted as 'ten to the power eleven'. The above value for the diameter of an atom would be written as 3.0×10^{-10} m, where $10^{-10} = 1 \div 10^{10} = 0.0000000001$. Note that a negative power of ten is a code for expressing powers of ten less than 1. Thus 10^{-10} is 1 divided by 'ten to the power ten', usually quoted as 'ten to the power minus ten'.

Scientific prefixes are used to represent certain powers of ten. For example, the prefix kilo- represents a thousand ($= 10^3$). Other standard scientific prefixes are listed below.

Prefix	Tera	Giga	Mega	kilo	milli	micro	nano	pico
Prefix symbol	T	G	M	k	m	μ	n	p
Power of ten	10^{12}	10^9	10^6	10^3	10^{-3}	10^{-6}	10^{-9}	10^{-12}

Notes

1 *The prefix symbol for micro, μ, is pronounced 'mu'.*
2 *1 centimetre (cm) = 10^{-2} m and 1000 grams = 1 kilogram.*

QUESTIONS

1 *Write the following values in scientific form in the same unit.*
   (a) 15 000 m,   (b) 0.000045 m,   (c) 650 000 000 kg,
   (d) 0.0000000035 kg

2 *Write the following values in scientific form in the required unit.*
   (a) *75 mm in metres,*   (b) *159 cm in metres,*
   (c) *56 000 km in metres,*   (d) *65 grams in kilograms,*
   (e) *0.027 µg in kilograms*

3 (a) *How many millimetres are there in a length of 1 metre?*
   (b) *How many milligrams are there in 1 kilogram?*

Density tests

Which is heavier, a kilogram of lead or a kilogram of feathers? This is a trick question because the weight of a kilogram of any substance on the Earth is the same, regardless of the substance. Anyone who answers that the lead is heavier than the feathers has made the mistake of thinking in terms of equal volumes of the two substances rather than equal masses. The point is that lead is much more dense than a sack of feathers.

The **density** of a substance is its **mass** per unit volume. The scientific unit of density is the kilogram per cubic metre (kg/m^3).

For example, the density of lead is about 11 000 kg/m^3. In comparison, the density of water is 1000 kg/m^3. Thus lead is 11 times more dense than water. The mass of a certain volume of lead is 11 times greater than the mass of the same volume of water.

Insight

The mass of a substance is a measure of the amount of matter the substance contains. Volume is the amount of space a substance takes up.

DENSITY CALCULATIONS

$$\text{Density} = \frac{\text{mass}}{\text{volume}}$$

The unit of density is the kilogram per cubic metre (kg/m^3).

The density formula can be rearranged as:

$$mass = volume \times density, \text{ or } volume = \frac{mass}{density}$$

Worked example

Calculate the mass of a volume of 20 m^3 of water. The density of water = 1000 kg/m^3

SOLUTION

Mass = volume × density = 20 × 1000 = 20 000 kg

DENSITY MEASUREMENTS – LIQUIDS

▶ *The volume of a liquid is measured by pouring the liquid into an empty measuring cylinder and measuring the level of the liquid in the cylinder against its graduated scale. The scale is usually marked in cubic centimetres (cm^3). To convert the reading into cubic metres (m^3), divide the measurement in cm^3 by 1 million (= 10^6) because 1 m^3 = 10^6 cm^3.*

Figure 1.3 Using a measuring cylinder.

▶ *The mass of a measured volume of liquid can be measured by using a top pan balance. This is used to measure the mass of an empty beaker. Then the liquid is poured into the beaker and the total mass of the beaker and the liquid is remeasured.*

The mass of the liquid is the difference between the total mass of the beaker and liquid and the mass of the empty beaker. A top pan balance usually gives a reading in grams (g). Note that 1000 g = 1 kg.

Figure 1.4 Using a top pan balance.

▶ *The density is calculated by dividing the mass in kilograms by the volume in m³.*

Sample measurements
Volume of liquid in measuring cylinder = 90 cm³ = 90 × 10⁻⁶ m³

Mass of empty beaker = 115 g

Mass of beaker and liquid from measuring cylinder = 220 g

Therefore, the mass of liquid = 220 − 115 = 105 g = 0.105 kg

$$\text{Density of liquid} = \frac{0.105 \text{ kg}}{90 \times 10^{-6} \text{ m}^3} = 1170 \text{ kg/m}^3$$

DENSITY MEASUREMENTS – SOLIDS

▶ *Use a top pan balance to measure the mass of a piece of a solid.*

- *Pour some water into a suitable measuring cylinder. Note the volume of the water from the reading of the water level. Tie the piece of solid to a thread and lower it into the water until it is completely submerged. The level of water in the cylinder will rise as a result. Note the new reading of the water level. The volume of the piece of solid is the difference between the two readings.*
- *The density is calculated by dividing the mass in kilograms by the volume in m^3.*

Notes

1 *The solid must be insoluble in water.*
2 *The volume of a regular solid such as a cube can be calculated from its dimensions. For example, the volume of a rectangular box is equal to its length × its height × its width.*

Sample measurements

Mass of object = 35.4 g = 0.0354 kg

Volume of water only in measuring cylinder = 50 cm^3

Volume of water and object in measuring cylinder = 63 cm^3

Therefore, the volume of the object = 63 – 50 = 13 cm^3 = 13 × 10^{-6} m^3

$$\text{Density of object} = \text{mass}/\text{volume} = \frac{0.0354 \text{ kg}}{13 \times 10^{-6} \text{ m}^3} = 2720 \text{ kg/m}^3$$

EUREKA!

Perhaps the most famous story in science is about Archimedes when he was asked by his king to find out if his new crown really was made of gold. The king thought the royal crown-maker might have cheated by making cavities in the gold which could have been filled with lead. He did not want to cut the new crown up, so he asked Archimedes to solve the problem without doing so. The solution came to Archimedes when he got into a bath of water and observed that the level of the water rose as he lowered himself into it. He realized that when an object is lowered completely into water, the displacement of the water is a measure of the volume of the object. According to legend, he greeted this discovery by running naked through the streets shouting 'Eureka!', which means 'I have found it'.

So how did Archimedes use this discovery to find out if the crown was a fake? Archimedes recognized that, if the crown was 100% gold, its weight in air divided by its volume should be the same as for any other gold object. So he weighed the crown and then measured its volume by measuring the volume of water displaced when it was immersed completely in water. He then calculated the weight of the crown in air, divided by its volume. Then he repeated the test and calculation with an object known to be made of gold. Fortunately for the royal crown-maker, the result was the same. Archimedes' measurements showed that the crown's density was the same as the density of a solid gold object. Gold is 18 times more dense than water. In effect, Archimedes discovered that the crown and a known gold object were both 18 times as dense as water. He therefore concluded that the crown was made of gold. Archimedes' test is the first recorded example of non-destructive testing.

Insight

Platinum is more dense than gold. Gold is the fourth most dense element. Compared with the atoms of much less dense elements, an atom of gold or platinum has much more mass but only a slightly bigger volume.

QUESTIONS

4 Calculate:
   (a) the mass of a volume of $5.0 \, m^3$ of sand of density $2500 \, kg/m^3$
   (b) the volume of a mass of 40 kg of steel of density $8000 \, kg/m^3$
   (c) the density of a brick of mass 6.0 kg and volume $0.002 \, m^3$.

5 A wooden cube of mass 0.80 kg measures $0.10 \, m \times 0.10 \, m \times 0.10 \, m$. Calculate: (a) its volume, (b) its density.

6 A glass pane has a height of 0.50 m, a width of 0.40 m and a thickness of 0.006 m.
   (a) Show that its volume is 0.0012 m³.
   (b) The density of the glass is 2600 kg/m³. Calculate the mass of this glass pane.

7 Air at atmospheric pressure and at room temperature has a density of 1.2 kg/m³. Calculate the mass of air in a room of dimensions 5.0 m long × 4.0 m wide × 3.0 m high.

8 Water has a density of 1000 kg/m³. Calculate the mass of water in a water tank of volume 6.0 m³ when the tank is full of water.

9 Calculate the density of a liquid in grams per cubic centimetre (g/cm³) from the following measurements:
   Volume of liquid = 56 cm³
   Mass of empty beaker = 95.3 g
   Mass of beaker and liquid = 147.6 g.

10 Calculate the density of a solid from the following measurements:
   Mass of solid = 48.8 g
   Volume of water in measuring cylinder without solid = 40.0 cm³
   Volume of water in measuring cylinder with solid immersed in water = 46.5 cm³.

10 THINGS TO REMEMBER

1   **Matter** *is composed of atoms.* **Atoms** *are too small to be seen individually.*

2   *A* **lever** *can be used to increase the effect of a force on an object.*

3   *If two objects are on a balanced beam pivoted at its centre of gravity, the heavier object is nearer the fulcrum.*

4   *For two weights* W_1 *and* W_0 *on a beam balanced at its centre of gravity,*

$$\frac{d_0}{d_1} = \frac{W_1}{W_0}$$

where d_1 *and* d_0 *are their distances from the fulcrum.*

5   *The* **SI system** *is the scientific system of units. Its base units include the metre, the kilogram and the second.*

6   *A numerical value in standard form is written as a number between 1 and 10 followed by the appropriate power of ten.*

7   *Common prefixes include M (mega) for a million, k (kilo) for a thousand, m (milli) for a thousandth and μ (micro) for a millionth.*

8   **Mass** *is a measure of the amount of matter in a substance.*

9   **Volume** *is the amount of space a substance takes up.*

10 **Density** *= mass/volume. The unit of density is the kilogram per cubic metre (kg/m³).*

2

Science in motion

In this chapter you will learn:

- *how Galileo took on the Church and established the foundations of modern science*
- *how to measure the motion of a falling object*
- *how to work out how fast objects move and how fast they accelerate or decelerate when they speed up or slow down.*

The Scientific Age in which we live began uneasily several centuries ago. Perhaps the discovery of the New World made some scientists and philosophers think about nature differently from those who accepted Aristotle's teachings. Galileo is generally considered to be the founder of modern science, as he demonstrated that Aristotle's way of doing science was fundamentally flawed because it ignored the crucial role of observations. Over succeeding centuries, Galileo's approach to science was developed and the link between laws and observations became evident as new discoveries were made, analysed and tested. The modern philosophy of science is that laws and theories hold true on the basis of never having been disproved. This philosophy was formulated by Sir Karl Popper, who realized that observations and measurements can never prove a theory or a law but, at any time, a single experiment is sufficient to disprove a theory or law.

Measuring motion

Galileo challenged the Church because his astronomical observations led him to conclude that the planets move round the Sun, not round the Earth. This heliocentric model was put forward a century before Galileo by Copernicus, a Polish-born medieval scholar whose interests included astronomy, economics, mathematics and medicine. Copernicus devoted many years of his life to studying ancient works to find out why objections to Ptolemy's model of the Universe had been rejected. He constructed his own version of Ptolemy's model and found it was necessary to have more than 30 spheres to account for all the observations of the motion of the stars and the planets. He realized such complexity was not necessary if the Sun, not the Earth, was at the centre, and he reduced the model to little more than a set of concentric circles representing the orbits of the planets. Copernicus feared that his revolutionary ideas would be ridiculed by his contemporaries and his major work, *De Revolutionibus Orbium Coelestium*, was not published until shortly before he died in 1543. It was presented as a mathematical challenge, rather than a new scientific theory, and it failed to have any immediate impact. However, in 1600, Giordano Bruno was burned at the stake by the established Church for promoting the Copernican model and his own view that the Universe is not finite, as implied by the Celestial Sphere model. Surprisingly, the Copernican model was not fully accepted by the Church until the early nineteenth century. In contrast, the modern Church is far more willing to listen and has organized scientific conferences on current developments in cosmology.

The change from the authoritarian and rigid theories of the medieval age to the experimental philosophy that underpins our Scientific Age may not have originated with Galileo, but there is little doubt that, through his discoveries, Galileo showed the power of science to change established views. Equally important, through his methods, he showed the significance of reliable and accurate observations. However, perhaps Galileo's most important contribution was to establish that science has limits, which should not be extended without valid evidence. In showing that experimental science should be used to explain what we observe, Galileo cut the link between physical science and religion that had existed for centuries. The link between biological science and religion continued well into the nineteenth century through the controversies surrounding Charles Darwin's theory of evolution. Even in this field, the modern science of genetics, which supports evolution, has been developed through scientific methods based on Galileo's approach to science.

GALILEO'S EXPERIMENTS ON MOTION

In a legendary demonstration at Pisa in Italy, Galileo was reported to have shown that two falling objects descend at the same rate, regardless of their weight. This could have been done by releasing two objects simultaneously from the top of the Leaning Tower and observing that they hit the ground at the same time, thus proving that they fall at the same rate.

Insight

Galileo's test with falling objects only works if air resistance is negligible. Release a piece of paper and a coin at the same time. The paper falls more slowly because the air resistance on it is not negligible.

Galileo wanted to find out how the time of descent would change if the distance fallen was changed. With falling objects, he was unable to measure the time of descent accurately. However, he realized that the motion of an object released at the top of an inclined plane was likely to be similar to that of a falling object but slow enough to time. He devised a water clock to measure the time taken by an object to reach certain distances down the plane. Pendulum clocks and other mechanical timing devices were inventions of the future. Galileo measured the amount of water running at a constant rate out of a tank from the moment the object was released at the top of the inclined plane to when it passed a certain marker. The amount of water collected thus served as a measure of the time of descent. You can repeat Galileo's experiment by using a stopwatch rather than a water clock to time a ball rolling down a flat inclined board.

1 *Start with zero incline and you will discover that the ball stops rolling after it has been given a push. The reason is that friction between the ball and the inclined plane makes the ball stop.*

2 *Adjust the incline just enough so the ball rolls along it at steady speed after it is given a push. Record the time taken by the ball to reach markers at equal distances down the incline. The measurements should show that the ball travels equal distances in equal times. This is an example of motion at* **constant speed**.

3 *Increase the incline so the ball rolls down the incline without being given an initial push. Record the time taken by the ball to reach equal distances down the incline. The measurements should show that the time the ball takes to pass from one marker to the next decreases as the ball moves down the slope. This is because the ball's speed increases as it moves down the slope. This is an example of* **accelerated motion**.

Galileo Galilei

Galileo was born in 1564 in Pisa, Italy, the same year of birth as William Shakespeare. As the son of a nobleman, Galileo was educated in a monastery and in 1595 became Professor of Mathematics at the University of Padua, one of Europe's leading universities at that time, in what was then the Republic of Venice. His Venetian paymasters allowed him to follow his own interests and his discoveries on motion would have been sufficient to win long-lasting recognition. From time to time, he made scientific instruments for commercial purposes such as measuring the density of precious metals and stones. In 1609, reports reached him about the invention of an optical device, the telescope, for making distant objects appear larger and closer. Within a short time, he had designed and constructed his own telescope, capable of making distant objects appear ten times larger and closer. He demonstrated the power of his telescope to the Senate, at the top of the

Campanile in Venice, by observing incoming ships 50 miles away, two or more days' sailing away from the port. Such information was very valuable to the merchants and brokers of Venice.

Galileo realized that he could use the telescope to study heavenly bodies. He wanted to gather evidence in support of the Copernican model and he devised a more powerful telescope, capable of magnifying objects 30 times. He was astounded when he first used this telescope to study the night sky. He observed ten times as many stars as can be seen directly, without a telescope. He found that the surface of the Moon is heavily cratered and he discovered the four innermost moons of Jupiter. Galileo's astronomical discoveries were widely reported in Europe. His Copernican views became known to the Church, who opened a file on him. Galileo hoped his observations and conclusions in support of the Copernican model would be accepted by the Church so that he would no longer need to rely on the protection of the Venetian authorities, who were anti-clerical. However, in 1613 he was reprimanded by the Church for his views and three years later he was banned from supporting the Copernican model. His attempts over the next decade to persuade the Church to support his views met with failure and so he decided to set out his ideas and support for the Copernican model in print in Italian, for all to read. He completed his work, *Dialogue on the Two Chief World Systems,* by 1630 and after some difficulties with the censor authorities it was published in Florence in 1632. Galileo's book was an instant bestseller and the Church reacted rapidly by banning it and stopping further reprints. Galileo, now in his seventieth year, was summoned to appear before the Holy Office of the Inquisition in Rome, on 12 April 1633. Galileo duly appeared in front of the 12 judges of the Inquisition. No consideration was given to the discoveries he had made. Instead, the judges sought to show that Galileo had breached the agreement of 1616 when he had been instructed not to hold or defend the Copernican view. Galileo's defence was that the ban imposed in 1616 did not prevent teaching the Copernican view. He argued that his *Dialogue* presented both views to the reader. The judges decided that Galileo had broken the 1616 ban and had acted deceitfully. He was threatened with torture, forced to recant and made to spend the rest of his life under house arrest at his home in Florence, where he died in 1642. Science as a high-profile activity came to a standstill in Catholic Europe for many years. However, Galileo's *Dialogue* and his discoveries were taken up vigorously by scientists in Northern Europe, where the Church had much less authority. The publicity attracted by Galileo's trial perhaps helped to promote the Copernican system. Galileo's immense contribution to the development of science cannot be understated.

MOTION IN A STRAIGHT LINE AT CONSTANT SPEED

Speed is defined as distance travelled per unit of time. The scientific unit of speed is the metre per second (abbreviated as m/s or ms^{-1}).

For an object moving at constant speed, the distance travelled by the object in a certain time can be calculated by multiplying the speed by the time taken.

$$\text{Distance} = \text{speed} \times \text{time taken}$$

For example, suppose a car is travelling at a constant speed of 12 m/s. The car therefore travels a distance of 12 metres each second and would travel a distance of 720 metres in 60 seconds (= 12 m/s × 60 seconds). In 1 hour, which is 3600 seconds, the car would travel a distance of 43 200 metres (= 12 m/s × 3600 seconds), or 43.2 kilometres.

For an object moving at varying speed, its average speed over a certain distance or time is defined as the distance travelled/the time taken.

$$\text{Average speed} = \frac{\text{distance travelled}}{\text{time taken}}$$

Worked example

A train leaves a station at 11.00 a.m. and arrives at the next station, 15 kilometres away, at 11.22 a.m. Calculate: (a) the time taken in seconds and (b) the average speed of the train in metres per second on this part of its journey.

SOLUTION

(a) Time taken = 22 minutes = 22 × 60 s = 1320 s.

(b) Average speed = distance/time taken = 15 000 m/1320 s = 11.4 m/s.

Miles and kilometres per hour

The speed limit on motorways is different in different countries. In the United Kingdom, it is 70 miles per hour.

▶ 1 mile = 1.6 kilometres, so 70 miles = 70 × 1.6 = 112 kilometres. Thus a speed of 70 miles per hour = 112 kilometres per hour.

▶ 1 kilometre = 1000 metres and 1 hour = 3600 seconds, so a speed of 1 kilometre per hour = 1000 metres/3600 seconds = 0.278 m/s. Thus a speed of 112 kilometres per hour = 31.1 m/s (= 112 × 0.278 m/s).

▶ Abbreviations: mph means 'miles per hour'; kph or km h^{-1} means 'kilometres per hour'.

▶ Conversions: 5 mph = 8 km h^{-1} = 2.2 m/s.

DISTANCE V. TIME GRAPHS

The motion of an object can be represented by a graph of distance, on the vertical scale (called the y-axis), against time, on the horizontal scale (called the x-axis). For an object moving at constant speed in a straight line, the distance moved by the object changes by equal amounts in equal times. Hence the graph of distance v. time for such an object is a straight line, as in Figure 2.1.

The steepness of the line depends on the speed of the object. In Figure 2.1, the steepness of the line is constant, which means that the speed of the object is constant. The greater the speed of the object, the greater the distance the object moves each second, and so the steeper the line is. The steepness of the line is called its **gradient** and is defined as the change of the quantity plotted on the y-axis divided by the corresponding change of the quantity plotted on the x-axis.

Thus the gradient of the line = change of distance/time taken = the speed of the object.

Insight

The steepness or gradient of a staircase is given by the rise from one step to the next, divided by the width of the step. Builders refer to the steepness of a staircase as 'the rise over the tread'.

Worked example

(a) Determine the speed of the object whose motion is represented by the line in Figure 2.1.

Figure 2.1 A distance v. time graph.

(b) How far would this object travel at this speed in (i) 1 minute, (ii) 30 minutes?

SOLUTION

(a) The gradient of the line = 20 metres/4 seconds = 5 m/s.

(b) (i) Distance moved in 1 minute = speed × time = 5 m/s × 60 s = 300 m.

(ii) Distance moved in 30 minutes = speed × time = 5 m/s × 30 × 60 s = 9000 m.

Equations and symbols

Because word equations such as 'distance = speed × time' become tedious and time-consuming, letters are used as 'shorthand' symbols for physical quantities. So 'distance = speed × time' is usually written as $s = vt$, where s represents distance moved, v represents speed and t represents time taken. Each symbol stands for a certain physical quantity, in accordance with an agreed scientific convention.

An equation may need to be rearranged by moving the symbols from one side to the other. The basic rule to remember is to do the same to both sides of the equation.

For example, in the equation $s = vt$, the symbol s is said to be the subject of the equation because it appears first as you read the equation from left to right. To make v the subject of the equation,

▶ *divide both sides of $s = vt$ by t, to give* $\dfrac{s}{t} = \dfrac{vt}{t}$

▶ *cancel t from the top and bottom of the right-hand side of the equation to give* $\dfrac{s}{t} = v$

▶ *swap the two sides of the equation over so v is read first;* $v = \dfrac{s}{t}$

VELOCITY AND SPEED

Velocity is defined as speed in a specified direction. The unit of velocity is the metre per second (m/s), the same as the unit of speed.

Two objects travelling at 60 mph in opposite directions along a motorway have the same speed but not the same velocity. An object moving along a circular path at constant speed has a continually changing direction of motion, so its velocity continually changes even though its speed is constant.

Insight

The distinction between speed and velocity is important if the direction of motion of an object changes.

Summary

Speed is distance travelled per unit time.

Velocity is speed in a given direction.

For an object moving at constant speed, the distance moved in a certain time = speed × time.

5 mph = 8 km/hour (km h^{-1}) = 2.2 m/s.

QUESTIONS

1 Calculate the distance moved:
   (a) in 1 minute by a car travelling at a constant speed of 12 m/s
   (b) in 30 minutes by a plane travelling at a constant speed of 250 m/s
   (c) in 10 minutes by a car travelling at a constant speed of 30 m/s.

2 Calculate the time taken for:
   (a) a train moving at a constant speed of 15 m/s to travel a distance of 1 kilometre
   (b) a car travelling at a constant speed of 20 m/s to travel a distance of 500 m

(c) a plane travelling at a constant speed of 150 m/s to travel a distance of 250 km.

3 Calculate the average speed of:
(a) a train that takes 20 minutes to travel a distance of 15 km
(b) an athlete who takes 10 s to run a distance of 100 m
(c) a plane that takes 1 hour 40 minutes to travel a distance of 800 km.

4 Convert the following speeds into m/s:
(a) 30 mph,   (b) 700 mph,   (c) 100 km/h,   (d) 12 km/min.

5 Two cars join a motorway at the same junction at the same time, travelling in the same direction. One of the cars travels at 30 m/s for 40 minutes, then stops at a service station for 20 minutes before rejoining the motorway just as the other car is passing the service station.
(a) Calculate the distance travelled by the first car from the junction where it joined the motorway to the service station.
(b) Show that the average speed of the other car was 20 m/s.

Accelerated motion along a straight line

Figure 2.2 A speed v. time graph for constant acceleration.

Acceleration is defined as change of velocity per unit of time. The unit of acceleration is the metre per second per second, usually written as m/s².

For an object moving in one direction along a straight line, its acceleration is constant if its speed changes by equal amounts in equal times. Suppose the speed of such an object changes from u to v in time t. This change is shown by the straight line on the speed–time graph in Figure 2.2.

The change of speed in time $t = v - u$

Therefore, the acceleration, $a = \dfrac{\text{change of speed}}{\text{time taken}} = \dfrac{v - u}{t}$

Note that a negative value of acceleration is referred to as **deceleration**. This is the term used for slowing down.

Worked example
A car accelerates, without change of direction, from rest to a speed of 30 m/s in 60 s. Calculate its acceleration.

SOLUTION
Initial speed $u = 0$, final speed $v = 30$ m/s, time taken $t = 60$ s

Therefore $a = \dfrac{v - u}{t} = \dfrac{30 - 0}{60} = 0.5$ m/s²

MORE ABOUT SPEED V. TIME GRAPHS

1 *The acceleration of an object = the gradient of the line on a speed v. time graph.*

The gradient of a line on a graph is a measure of its steepness. In Figure 2.2, the triangle underneath the line has a height corresponding to the change of speed $v - u$, and a base corresponding to the time taken t. The gradient of the line is given by the triangle's height/its base. Since acceleration = change of speed/time taken = $(v - u)/t$, the acceleration is given by the gradient of the line.

2 *The distance travelled = the area between the line and the x-axis.*

Figure 2.3 shows that a train moves from A to B at a constant speed of 20 m/s in 250 s. The distance it travels from A to B is therefore 5000 m (= speed × time = 20 m/s × 250 s). This is represented by the area of the graph under the line AB. We can use this 'area' method on any speed v. time graph to find the distance moved. For example, from O to A in Figure 2.3, the area under the line is a triangle with a height and base corresponding to 20 m/s and 100 s respectively. Since the area of the triangle is ½ × its height × its base = ½ × 20 m/s × 100 s (= 1000 m), the distance travelled from O to A is therefore 1000 m.

Note that, in the above calculation, ½ × 20 m/s corresponds to the average speed of the train between O and A. Multiplying this average speed by the time taken gives the distance travelled. Since the average speed = ½ $(u + v)$, the distance travelled in time t is given by s = the average speed × the time taken = ½ $(u + v)$ t.

Worked example
A train on a straight line accelerates from rest to a speed of 20 m/s in a time of 100 s, then travels at constant speed for 250 s. It then decelerates at a constant rate for 50 s and stops.

(a) Draw a speed v. time graph to represent the motion of the train.

(b) Calculate the acceleration in (i) the first 100 s, (ii) the next 250 s, (iii) the final 50 s.

(c) The train travelled a distance of 1000 m from O to A and 5000 m from A to B. Calculate (i) the distance travelled from B to C, (ii) the average speed of the train over the journey.

SOLUTION
(a) See Figure 2.3.

Figure 2.3 Speed v. time graph for a moving train.

(b) (i) 1st stage (0 to 100 s); $u = 0$, $v = 20$ m/s, $t = 100$ s

Therefore, acceleration, $a = \dfrac{v - u}{t} = \dfrac{20 \text{ m/s} - 0}{100 \text{ s}} = 0.2$ m/s^2

(ii) 2nd stage (100 to 350 s); $u = v = 20$ m/s, $t = 250$ s

$a = 0$ because there is no change of its speed

(iii) 3rd stage (250 to 300 s); $u = 20$ m/s, $v = 0$, $t = 50$ s

$a = \dfrac{v - u}{t} = \dfrac{0 - 20 \text{ m/s}}{50 \text{ s}} = -0.4$ m/s^2

(c) (i) Distance travelled from B to C = ½ × height × base of the triangle under BCE = ½ × 20 m/s × 50 s = 500 m

(ii) Total distance travelled = 1000 + 5000 + 500 = 6500 m

Total time taken = 100 + 250 + 50 s = 400 s

Therefore the average speed $= \dfrac{\text{total distance}}{\text{time taken}} = \dfrac{6500 \text{ m}}{400 \text{ s}} = 16(.25) \text{ m/s}$

GALILEO AND GRAVITY

Galileo showed that an object rolling down an inclined plane gathered speed at a constant rate as it descended. He showed that the acceleration was constant and could be made larger by increasing the steepness of the incline.

He thought that the acceleration of an object falling freely ought to be constant and ought to be the same for any object. However, he was unable to time the vertical descent of an object because his clock was not accurate enough.

One method of timing the descent of a falling object is to photograph a small steel ball as it falls in front of a vertical scale. If the apparatus is in a darkened room that is illuminated by a flashing stroboscope, an image of the ball is captured on the photograph each time the light flashes. Provided the light flashes at a constant rate, the photograph therefore shows the position of the ball at equal intervals. A video camera could be used instead of a film camera but the stroboscope would still be needed. If the flash rate of the stroboscope is known, the time interval between successive flashes can be worked out. Figure 2.4 shows the idea. Measurements from such an investigation are shown below.

Time from start (s)	0.0	0.1	0.2	0.3	0.4	0.5
Distance fallen (m)	0.0	0.05	0.20	0.45	0.80	1.25
Average speed (m/s)	0.0	0.50	1.00	1.50	2.0	2.5
Speed (m/s)	0.0	1.00	2.00	3.00	4.00	5.00

Figure 2.4 Investigating free fall.

The average speed between the start and each flash has been worked out by dividing the distance fallen by the time taken from the start. Because the initial speed is zero, the speed at each flash is twice the average speed. The results show that the speed increases by 1.0 m/s every 0.1 s, which scales up to an increase of 10 m/s every second. Thus the acceleration of free fall has a constant value of 10 m/s^2. This value is usually referred to as g, the acceleration due to gravity of a freely falling object. Accurate measurement of g shows that g varies slightly with latitude. Its value is 9.81 m/s^2 at the Earth's poles and 9.78 m/s^2 at the Equator. The acceleration due to gravity $g = 9.8$ m/s^2 at the Earth's surface.

Insight

The European Space Agency's Gravity and Ocean Circulation Explorer (GOCE) satellite, launched in March 2009, will make accurate measurements of g across the entire surface of the Earth, enabling the precise shape of the planet to be determined. This will help scientists to learn more about climate change and about the Earth's crust. Use your search engine to find out more about ESA's GOCE satellite.

QUESTIONS

Assume $g = 9.8 \, m/s^2$

6 A car accelerates from rest at an acceleration of $0.5 \, m/s^2$ for 20 s. Calculate:
   (a) its speed after 20 seconds from rest,
   (b) the distance it moves in this time.

7 In an experiment, a ball was released from rest at the top of a slope and rolled a distance of 2.0 m down the slope in 8.0 s. Calculate:
   (a) its average speed in this time,
   (b) the speed after 8.0 seconds from rest,
   (c) the acceleration of the ball.

8 An aircraft accelerates from rest to a speed of 120 m/s in 40 s when it takes off. Calculate:
   (a) its average speed during take off,
   (b) the distance it travels in this time,
   (c) its acceleration during this time.

9 A small object is released from the top of a well and hits the water surface of the well 1.5 s later. Calculate:
   (a) its speed just before it hits the water surface,
   (b) the distance from the top of the well to the water surface.

10 A rocket was launched vertically and accelerated at a constant rate for 50 s, from the moment it was launched to when its fuel ran out. At this point its speed had reached 200 m/s.
   (a) Calculate the acceleration of the rocket during this time.
   (b) Calculate the height of the rocket when its fuel ran out.
   (c) Explain why the rocket continued to gain height for a further 20 s (to the nearest second) after its fuel ran out.
   (d) Calculate the maximum height attained by the rocket.

10 THINGS TO REMEMBER

1 **Speed** *is distance travelled per unit time.*

2 **Velocity** *is speed in a given direction.*

3 *The unit of speed and of velocity is the metre per second (m/s).*

4 **Acceleration** *is change of velocity per unit time.*

5 *The unit of acceleration is the metre per second per second (m/s²).*

6 *The gradient of a distance v. time graph represents speed.*

7 *The gradient of a speed v. time graph represents acceleration.*

8 *The area under a speed v. time graph represents distance travelled.*

9 *The acceleration of free fall is about 10 m/s².*

10 *For an object that undergoes constant acceleration a without change of direction from initial speed u to speed v after time t,*

$$acceleration, a = (v - u)/t$$
$$distance\ travelled, s = \tfrac{1}{2}\ (u + v)t$$

3

Forces in action

In this chapter you will learn:

- *how Newton in the late seventeenth century formed the principles of physics for the next two centuries*
- *the difference between mass and weight*
- *how to lose weight without much effort*
- *how to use Newton's ideas in everyday situations, such as making sure objects do not fall over.*

In the two centuries after Galileo's death, science developed at an ever-increasing rate. By the end of the nineteenth century, many scientists believed that all the fundamental principles of physics had been discovered and little else remained except to make more accurate measurements of the properties of matter. A few niggling issues remained that could not be properly explained, mostly in connection with light, but it was generally thought that the problems would be solved using the known principles of physics. The laws of mechanics had been established by Sir Isaac Newton in the late seventeenth century and had proved enormously successful in explaining the motion of objects, from the throw of a ball to the motion of the planets in orbit round the Sun. Michael Faraday had discovered how to generate electricity in the 1820s and, by the end of the century, his ideas led to the establishment of the electricity supply industry.

The foundations of physics that seemed so secure and certain at the end of the nineteenth century fell apart in the first two decades of the twentieth century. We will look at how and why this happened later in this book. Einstein's theories of relativity and the quantum theory reshaped the principles of the subject and new discoveries about the properties of matter have resulted in major new industries such as electronics, nuclear power and aerospace. However, pre-twentieth century physics, which is usually referred to as classical physics, arguably had an even greater impact because the principles of classical physics underpinned the great developments in engineering and construction that created the industrial revolution of the nineteenth century. In this chapter we will consider in detail the laws of motion established by Newton and how these classical laws are used in some everyday situations.

Newton's laws of motion

Galileo's ideas on motion and the Copernican system were brought together by Sir Isaac Newton in his laws of motion and **gravitation**. For over two centuries, these laws were thought to be universal, applicable in any situation where material bodies interact. Newton's three laws of motion underpin much of modern technology and engineering. The motion of planets, comets and satellites can be explained almost completely using Newton's theory of gravitation and his laws of motion.

Before Newton wrote his laws of motion, he defined key physical quantities which he then linked in his laws of motion to each other and to other physical quantities already defined.

1 *The quantity of matter in an object, or its* **mass,** *was defined by Newton as its volume × the density of the substance from which the object is composed. For example, iron is about 8 times as dense as water, so the mass of 2 cubic metres of steel is 16 times the mass of 1 cubic metre of water. The mass of an object is now determined by comparing its weight with the weight of a standard mass.* See p. 7.

2 *The quantity of motion of a moving object, or its* **momentum,** *was defined by Newton as its mass × its velocity. For example, the momentum of an object of 2 kg mass moving at 3 m/s is 6 times the momentum of an object of mass 1 kg moving at 1 m/s.*

3 *A force is an action on an object that changes its motion. Two or more forces acting on an object balance each other out if the object stays at rest or moves at constant speed along a straight line.*

Insight

We say that the motion of an object is uniform, if its velocity is constant – in other words, if its speed and direction of motion do not change.

NEWTON'S 1ST LAW OF MOTION

An object continues at rest or in uniform motion unless acted on by a force.

Here are some everyday situations that reveal the link between force and motion as established by Newton in his 1st Law:

1 *An object sliding across ice moves at constant speed without change of direction because, due to the ice, there is no force acting on the object. If the object crosses a surface where there is no ice, it is brought to a standstill by the force of friction between the object and the surface.*

2 *A cyclist in motion on a level road gradually comes to a halt if he or she stops cycling and 'freewheels'. The reason is that friction in the wheel bearings and air resistance on the cyclist act against the motion and gradually stop the cycle moving.*

3 *An object being whirled round on the end of a length of string is pulled by the string so its direction repeatedly changes. If the string suddenly snaps, the object flies off at a tangent because the pull force of the string has suddenly stopped acting on the object.*

In effect, Newton's 1st Law tells us what a force is – namely, anything that can change the motion of an object. Different types of force include:

- ▶ *the force of gravity on an object, usually referred to as the weight of the object*
- ▶ *the force of friction acting on two surfaces when they slide over each other*
- ▶ *the tension in an object such as a string when it is taut and being pulled at each end*
- ▶ *the force on an object when another object pushes or pulls it or supports it*
- ▶ *the force between two electrically charged objects or between two magnetized objects.*

Balanced forces

An object acted on by two or more forces can be at rest or in uniform motion if the forces on it balance each other out. The simplest example of this situation is a tug-of-war when a stalemate exists, because the two teams pull on the rope with equal force in opposite directions.

force F_1 force F_2

In a tug-of-war 'stalemate', the teams pull with equal and opposite forces. The forces balance each other out.

$$F_1 = F_2$$

tension T

The weight of an object hanging on the end of a vertical rope is equal and opposite to the tension in the rope.

$$T = W$$

weight W

Figure 3.1 Balanced forces.

Another example is when a rope hanging vertically supports a bucket. The force of gravity on the bucket acts downwards and is equal and opposite to the tension in the rope which acts upwards, as shown in Figure 3.1.

NEWTON'S 2ND LAW OF MOTION

The force on an object is proportional to the rate of change of momentum of the object.

An object initially at rest and acted on by a single force gains speed and therefore gains momentum. For a given amount of force, the object gains a certain amount of momentum every second during the time the force acts on it. Consider the example of an object of mass m being accelerated along a straight line by a constant force F, acting in the same direction as the object is moving, for a time t. Suppose the object's speed increases from u to v in this time, without change of direction.

Force and motion

initial speed = u speed = v at time t

Figure 3.2 Newton's 2nd Law of motion.

▶ The initial momentum of the object = mu
▶ The final momentum of the object = mv

so the change of momentum = final momentum – initial momentum:

$$mv - mu$$

and the rate of change of momentum = $\dfrac{\text{change of momentum}}{\text{time taken}}$:

$$\frac{mv - mu}{t} = \frac{m(v - u)}{t}$$

According to Newton's 2nd Law, the force F on the object is proportional to the rate of change of momentum.

Hence:

$F = km\dfrac{(v-u)}{t}$ where k is a constant of proportionality.

Because the acceleration of the object $a = \dfrac{(v-u)}{t}$, then $F = kma$.

By defining the unit of force, the **newton**, as the amount of force that would give an object of mass 1 kg an acceleration of 1 m/s², the value of k is therefore set at 1. Newton's 2nd Law can then be written as the equation

$$F = ma$$

where F = force in newtons, m = mass in kilograms, and a = acceleration in m/s².

Worked example

A car of mass 600 kg accelerates from rest to a speed of 12 m/s in 30 s. Calculate (a) its acceleration, (b) the force needed to produce this acceleration.

SOLUTION

Initial speed, $u = 0$, final speed $v = 12$ m/s, time taken, $t = 30$ s

Therefore, $a = \dfrac{(v-u)}{t} = \dfrac{12-0}{30} = 0.4$ m/s²

$F = ma = 600$ kg $\times$ 0.4 m/s² = 240 N.

Insight

If the force acts in the opposite direction to the object's direction of motion, the object decelerates.

Here are some simple experiments to demonstrate Newton's first two laws of motion.

1 **Newton's 1st Law of motion:** Flick a coin across a very smooth level surface (e.g. a tea tray) and it will slide across the surface. In the absence of friction, the coin continues to move without being pushed because no force acts on it.

2 **Newton's 2nd Law of motion:** Release a coin at the top of a flat slope (e.g. a tea tray propped up at one end) so that it slides down the slope. Observe it gather speed as it moves down the slope. Its momentum increases because its weight provides a steady force which pulls it down the slope.

Sir Isaac Newton 1642–1727

The ideas about machine power were first developed by Sir Isaac Newton when he applied his ideas about force and motion to machines. Newton was born in the market town of Grantham in the county of Lincolnshire in eastern England. His father died before he was born and he was brought up by a grandparent after his mother remarried. He was sent to the local grammar school as a boarder and entered Cambridge University in 1661. England at this time was a republic under Oliver Cromwell. The University was closed at times during 1665 and 1666 because of the Great Plague which ravaged the country during those years. Newton returned to his home in the Lincolnshire countryside and in just two years produced mathematical theorems and physical theories that revolutionized mathematics and physics. He returned to Cambridge in 1667 and was appointed two years later, at 26 years of age, to the Chair of Mathematics at Trinity College.

He published his theories of mathematics and physics in 1687 in his greatest work, *Philosophiae Naturalis Principia Mathematica* (The Principia), in which he showed that his three laws of motion and his law of gravitation are sufficient to explain the motion of any system of bodies. He proved once and for all that the planets and the Earth go round the Sun and he explained Kepler's laws of planetary motion and Galileo's observations on the motion of falling objects. Using his law of gravitation, which we will meet in Chapter 13, he was able to predict comets, eclipses and tides. He saw the Universe as a gigantic mechanical system, governed by the same laws that govern the motion of objects on the Earth. His ideas provided the guiding principles for science for the next two centuries, until Einstein showed that space and time are not independent quantities. We will meet Einstein's ideas in detail in Chapter 10.

Newton's interests in science were wide-ranging and included astronomy, chemistry and optics as well as mathematics and physics. He left the University in 1696 to become Master of the Mint where he devoted his talents to monetary reform, only occasionally delving back to the challenges of science. His pre-eminence as a scientist was recognized in 1703 when he was elected as President of the Royal Society, and he was knighted in 1705. In contrast with Galileo, who was shunned by the Catholic Church, Newton became part of the Establishment in England, even attracting the attention of satirists, who were clearly unaware that the freedom of thought on which they depended would not have emerged without Galileo and Newton.

Mass and weight

The acceleration of a freely falling object, g, is constant, provided the distance fallen is much smaller than the radius of the Earth. This acceleration is caused by the weight of the object, which is the force of gravity pulling it towards the Earth. Using Newton's 2nd Law, in the form $F = ma$, therefore gives mg as the force of gravity on an object of mass m. Hence the weight W of an object of mass $m = mg$.

Weight W (in newtons) = mg, where m = mass in kilograms and g = the acceleration of a freely falling object (= 9.8 m/s^2 near the Earth's surface).

The weight of a 1 kg mass near the Earth's surface is therefore 9.8 N. The weight of a person of mass 60 kg at the Earth's surface is 588 N.

The **gravitational field strength** at any position is defined as the force of gravity per unit mass on an object at that position. Hence gravitational field strength = $mg/m = g$. Thus g can be referred to as either the acceleration of a freely falling object (in m/s^2) or gravitational field strength (in newtons per kilogram).

NEWTON'S 3RD LAW

When two bodies interact, they exert equal and opposite forces on each other.

For example, if you clap your hands together, the force of your left hand on your right hand is equal and opposite to the force of your right hand on your left hand.

If you lean on a wall, the wall exerts a force on you that is equal and opposite to the force you exert on the wall. When you stand still, the floor exerts an upward force on you that is equal and opposite to the force you exert on the floor.

Insight

The Earth exerts a force on you and you exert an equal and opposite force on the Earth. The force is due to the gravitational attraction between you and the Earth.

QUESTIONS

$g = 9.8 \, m/s^2$

1   Explain why a car is likely to skid at a bend on an icy road.

2   (a) Calculate the weight of an object of mass (i) 2 kg, (ii) 4 kg.
     (b) Explain why a 2 kg object and a 4 kg object released above the
          ground at the same time fall at the same rate.

3   (a) A bucket of mass 2.5 kg contains 0.008 m^3 of water.
          (i) Calculate the mass of water in the bucket. The density of
               water is 1000 kg m^{-3}.

(ii) Calculate the total weight of the bucket and the water.
(b) The bucket is attached to a rope, which is used to raise the bucket. Calculate the tension in the rope when it supports the bucket at rest containing $0.008 \, m^3$ of water.

4 Calculate:
(a) the force needed to give an object of mass 4 kg an acceleration of (i) $2 \, m/s^2$, (ii) $6 \, m/s^2$
(b) the acceleration of an object of mass 2 kg acted on by a force of (i) 10 N, (ii) 8 N.

5 A car of mass 800 kg is brought to rest from a speed of 30 m/s in a time of 15 s. Calculate:
(a) the deceleration of the car
(b) the force needed to produce this deceleration.

Stability

FORCE AS A VECTOR

A physical quantity that has a direction as well as a size is a **vector** quantity. Examples of vectors include force, velocity, acceleration and momentum. A physical quantity that is not directional is referred to as a **scalar** quantity. Examples of scalars include distance, speed, mass and energy.

A vector quantity may be represented by an arrow of length in proportion to the quantity and pointing in the appropriate direction. For example, a force of 10 N acting horizontally due North on an object may be represented on a diagram as an arrow of length 10 cm pointing in a direction defined on the diagram as due North. Caution is needed to use Newton's 3rd Law correctly when force diagrams are drawn. For every force acting on a body, there is an equal and opposite force exerted by the body. To ensure that force diagrams do not become complicated, a force diagram for an object should show only the forces acting on the body, not the forces exerted by the body.

CENTRE OF GRAVITY

When an object is acted upon by two or more forces including the force of gravity, the effect on the object depends on the direction and size of the forces. Each force acting on an object may be represented as a vector on a force diagram. To represent the force of gravity on an object (i.e. its weight), which acts on all parts of the object, we define the centre of gravity of an object as the point where its entire weight may be considered to act. The weight of an object can then be shown on a force diagram as a vector arrow pointing downwards at the centre of gravity.

The idea of centre of gravity, sometimes referred to as centre of mass, enables us to understand stability and the conditions necessary to keep an object in equilibrium. The examples below are chosen to illustrate the idea.

1 *A tall free-standing bookcase is liable to topple over when it is moved, unless it is supported. If it tilts too much, its centre of gravity moves outside its base and it topples over, unless*

Figure 3.3 Toppling.

someone supports it to prevent this. Push a tall object such as an empty cereal packet sideways at its top and you will find that it tilts, then topples over if it is pushed too far. If you hold it at the point where it only just topples, you should find the centre of the box is directly above the edge in contact with the table.

Insight

High-sided vehicles can topple over in strong winds. The force of a side wind on such a vehicle causes a turning effect, which would make the vehicle difficult to steer and could cause it to topple over.

2 *A beam supported on a pivot that is not at the centre of the beam can be balanced by hanging a single weight from the beam between the pivot and the end of the beam nearest the pivot. Figure 3.4 shows the arrangement.*
The centre of gravity of a uniform beam is at its centre. Using the principle of the lever (see p. 5)

$$W_1 d_1 = W_0 d_0$$

where W_1 is the weight of the hanging object, W_0 is the weight of the beam, d_1 is the distance from W_1 to the pivot and d_0 is the distance from the pivot to the centre of gravity of the beam.

Figure 3.4 Balancing a beam off-centre.

Note that d_0 is the distance from the pivot to the centre of gravity of the beam. Hence the weight of the beam W_0 can be calculated if W_1, d_0 and d_1 are known.

flat card
freely suspended
from rod

x

rod through
hole X in card

y

Centre of
gravity directly
below rod

z

plumb
line

vertical line
marked on card
from plumb line

Note: repeat test with the card
suspended at Y and Z in turn.
The centre of gravity of the card
is where the lines intersect.

Figure 3.5 Finding the centre of gravity of a flat object.

Locate the centre of gravity of a flat object

Cut a piece of card to make a large triangle. Pierce a small hole
near one corner of the triangle and hang it freely from a pin.
Hang a 'plumb line', consisting of a cotton thread supporting
a small weight, from the pin against the triangle. Draw a pencil
line on the triangle along the line of the thread. When the triangle
is at rest, its centre of gravity is directly below the pin along this
line. Repeat the test by hanging the triangle from a pinhole at
each of the other two corners in turn. The centre of gravity of
the card triangle is where the lines through the pinholes intersect.

QUESTIONS

6 Sketch a free body force diagram for:
   (a) a bucket at rest hanging on the end of a vertical rope
   (b) a tightrope walker balanced on a tightrope.

7 (a) Explain why a free-standing bookcase topples over easily if
        only the top shelf is filled with books.
   (b) Explain why a person wearing a heavy backpack is liable to
        topple backwards when sitting on a wall.

8   Explain why a full bottle of lemonade is less stable than a bottle that is half-full.

9   A child's seesaw consists of a uniform beam of length 4.0 m pivoted at its centre. The seesaw is balanced when a girl sits at one end and a boy of weight 200 N sits on the seesaw 1.0 m from the other end.
    (a) Sketch the force diagram for the seesaw.
    (b) Calculate the weight of the girl.

10 A uniform beam of length 5.0 m is positioned at right angles to the edge of a river bank so that a 2.0 m section of the beam projects from the bank.
    (a) Explain why the beam does not fall into the river in this position.
    (b) Why would the beam topple into the river if someone attempted to walk on it beyond a certain distance from the bank?

10 THINGS TO REMEMBER

1 **Momentum** = *mass × velocity.*

2 *Newton's first law of motion: An object continues at rest or in uniform motion unless acted on by a force.*

3 *Newton's second law of motion: The force needed to accelerate an object is given by the equation* **force = mass × acceleration.**

4 *Newton's third law of motion: When two objects interact, they exert equal and opposite forces on each other.*

5 *The unit of force is the newton (N).*

6 **Weight** = *mass × g.*

7 *Directional quantities such as force and velocity are called* **vectors.**

8 *Non-directional quantities such as mass and speed are called* **scalars.**

9 *The* **centre of gravity** *of an object is the point where its entire weight may be considered to act.*

10 *The lower the centre of gravity of a free-standing object, the more stable it is.*

4

Machines at work

In this chapter you will learn:
- *about energy and how you can transfer it and use it*
- *what is meant by the Principle of Conservation of Energy*
- *what is meant by the efficiency of a machine and how it can be worked out*.

The industrial revolutions in the eighteenth and nineteenth centuries developed because engineers were able to make machines and engines that were more powerful than the windmills and watermills of the pre-industrial era. The scientific principles established by Newton enabled engineers to design and build such machines and engines. With Newton's guiding principles, engineers continue to seek more efficient and cost-effective solutions to modern problems. In this chapter we look at how the ideas established by Newton are used to explain how different types of machine work.

Work, energy and power

When a lever is used to shift an object, for example, to prise open a lid, a force applied by hand at the end of the lever has a much larger effect via the lever than if it was applied directly. The greater the distance from the pivot to the point where the force is applied, the greater the effect. A lever is a force multiplier. The effect of the applied force is multiplied, so a much larger force acts. Another example of a lever at work is a manual punch used to punch a hole in sheets of paper. When a downward force is applied by hand to the lever, the metal cutter is pushed through the paper. The force of the hand on the lever is multiplied tenfold because this force acts 10 times further from the pivot than the metal cutter.

Figure 4.1 A manual punch.

Although most devices used to move objects are more complicated than a simple lever, the same general principle applies. The device exerts a force to move an object (which we will refer to as the load) as a result of a force (which we shall refer to as the effort) applied to the device. The effort needed to shift a given load is smaller than the load, because of the way the device is designed. For example, a heavy weight supported by a rope attached to the drum of a winch can be raised by turning the handle of the winch. When the handle is turned, the rope winds round the drum, thus raising the weight attached to the rope. The effort applied to the handle acts at a much larger distance from the centre of the drum than the pull of the load on the rope on the drum. The pull of the rope on the drum is the load that has to be shifted by the effort. Figure 4.2 shows an end-view of the drum. The law of levers (see p. 5) can be applied here to give the equation

$$F_1 d_1 = W_2 d_2$$

where F_1 is the effort, d_1 is the distance from the line of action of the effort to the axis of the drum, W_2 is the weight to be raised and d_2 is the radius of the drum. Since d_1 is much greater than d_2, then F_1 is much smaller than W_2.

Figure 4.2 Using a winch.

Suppose that, in the above example, distance d_1 is ten times greater than distance d_2. A 500 N weight could therefore be raised by an effort of 50 N, provided no other forces were present. Each turn of the handle would raise the weight by a height equal to the circumference of the drum, which is equal to $2\pi d_2$. The effort is applied at a much greater distance d_1 from the axis. Therefore, in one turn, the effort goes round a circle of circumference equal to $2\pi d_1$ which is ten times greater than that of the drum. Thus the reduction of effort is at the expense of an increase of distance over which the effort must be applied. In other words, as the load is 10 times the effort, the effort must move a distance of 10 metres for every one metre the load is raised. If the winch handle is turned at a steady speed of 1 m/s, the weight would therefore rise at a steady speed of 0.1 m/s. More generally, for a given load,

the load × the load speed = the effort × the effort speed.

The above example shows that the **power** of any machine depends on the load it moves and the speed at which it moves the load. A machine that shifts four times as much load and does the job twice as fast as another machine must be eight times more powerful than the other machine.

▶ *The output power (in watts) of a machine in operation is the load (in newtons) × the speed of the load (in m/s). A machine shifting a load of 100 N at a speed of 10 m/s has an output power of 1000 watts, which is 1000 times more than the output power of a machine shifting a load of 1 N at a speed of 1 m/s.*

▶ *The power supplied to a machine is the effort × the speed at which the effort moves. For a machine that is 100% efficient, this is equal to the power output of the machine, because the load × the load speed = the effort × the effort speed. In other words, provided no power is wasted*

The power output of a machine = the power supplied to it.

Insight

Imagine two lifts ascending in a building. If one of them carries three times as many people as the other one at twice the speed, it must be six times more powerful.

WORK AND ENERGY

The **work** done by a force on an object is defined as the force × the distance moved by the object along the line of action of the force. The unit of work is the joule (J), which is equal to the work done when a force of one newton moves its point of application through a distance of one metre in the direction of the force.

$$\text{Work done (in joules)} = \text{force (in newtons)} \times \text{distance moved in the direction of the force (in metres)}$$

For example, the work done on a brick of weight 2.0 N lifted through a height of 1 m is equal to 2 J. If the same brick had been raised through only 0.5 m, the work done on it would have been 1 J.

Insight

The distance moved must be in the direction of the force. If an object is moved horizontally, its potential energy is unchanged because the force of gravity on it is vertical and the object does not move in a vertical direction.

The work done on an object by raising it enables the object to do work on something else when it drops. A brick dropped onto someone's shoe would do work in denting the shoe. The work done by the brick on the shoe could not be recovered, however, unless the shoe cap was elastic. A raised object has the capacity to do work because of its position. Another example is a clockwork spring which has work done on it when it is wound up. When the spring unwinds, it does work by making the wheels and gears inside the clock move. If the spring is prevented from unwinding, it keeps the capacity to do work until it is allowed to unwind. This capacity to do work is defined as the **energy** of an object. The term 'stored work' might seem more appropriate, but work is only done when a force moves, so the term 'energy' is used for the capacity of a body to do work.

> The energy of an object is its capacity to do work.

Energy is measured in joules, the same unit as the unit of work. For example, if a brick of weight 5 N is raised by 2 m from the ground, the work done to raise it is equal to 10 J (= 5 N × 2 m). The raised brick therefore contains an extra 10 J of energy due to being lifted. If the brick is dropped onto someone's shoe, it causes a dent in the shoe and loses the extra 10 J of energy in the process. Thus 10 J of work must have been done to create the dent. You can work out the force of the impact from the depth of the dent by recalling that work done = force × distance. For example, a dent of 0.5 cm (= 0.005 m) would correspond to a force of 2000 N (= 10 J / 0.005 m) and a really big 'ouch'!

Energy transformations
Objects can possess energy in various ways, referred to as 'forms' of energy. For example:

▶ **Kinetic energy** *is the energy of a moving object due to its motion. The faster an object moves, the more kinetic energy it has. An unbalanced force is needed to make an object move faster, so the work done by the force is equal to the gain of kinetic energy of the object.*

- ▶ **Potential energy** *is the energy of an object due to its position. For example, an object raised from the ground gains potential energy because a lift force opposite to the weight of the object is needed to raise the object.*
- ▶ **Elastic energy** *is the energy stored in an object when it is stretched or squeezed. When the object returns to its undistorted shape, the elastic energy is released.*
- ▶ **Electrical energy** *is the energy of particles such as electrons that carry electric charge. A capacitor is a device used to store electric charge. Energy is stored in this process, since work must be done to force electrons into a capacitor because identical charged particles repel each other.*
- ▶ **Thermal energy** *is the energy of an object due to its temperature. For example, in a car engine, the petrol–air mixture in a cylinder is ignited by means of a spark which causes the air to become very hot very quickly. The pressure of the hot air shoots up and forces the piston out so it turns the drive shaft. See p. 85.*
- ▶ **Nuclear energy** *is energy that can be released from an atom which has an unstable nucleus. See p. 212.*
- ▶ **Chemical energy** *is the energy that can be released when chemicals react. For example, when a battery is used to light a torch bulb, chemical energy is released inside the battery to force electricity through the torch bulb.*

Other forms of energy include sound energy and light energy. Heat transfer is energy transfer due to temperature difference. When a hot object is in contact with a cold object, heat transfer from the hot object to the cold object reduces the thermal energy of the hot object and increases the thermal energy of the cold object. See pp. 68–9.

Energy can change from one form into other forms. We say that energy is 'transformed' when it changes from one form into another. For example, an object released above the ground loses potential energy and gains kinetic energy as it falls. So its gravitational potential energy is transformed into kinetic energy. If it hits the ground without bouncing, its kinetic energy is transformed into thermal energy and sound energy on impact.

The Principle of Conservation of Energy

When energy is transformed or transferred between objects in an isolated system, the total energy of all the objects after the change is the same as the total energy of all the objects before the change. The total energy of all the objects is conserved (i.e. unchanged). This is known as the Principle of Conservation of Energy. Energy is the capacity to do work. The Principle of Conservation of Energy means that the total capacity (to do work) of an isolated system of objects is unchanged by interactions between the objects. The principle has been tested many times and the measurements always come up with the result that the total energy is unchanged. When energy transfer takes places within the system, individual objects do work on each other, so gaining or losing energy. However, the total capacity to do work is unchanged. In other words, the total energy is unchanged.

Some thoughtful experiments

▶ *A pendulum bob on a string swings from side to side repeatedly. The pendulum bob loses potential energy and gains kinetic energy when it moves towards the centre. When it moves away from the centre, it loses kinetic energy and gains potential energy. Its total energy stays the same as, at any point, its kinetic energy plus its potential energy always add up to the same.*

▶ *When an object crashes to the floor without rebounding, it loses all its kinetic energy. What happens to this energy? A test using a mass of lead pellets dropped down a long tube shows that the temperature of the pellets increases. The kinetic energy of the pellets becomes thermal energy.*

▶ *When a small weight is used to raise a large weight at a steady speed using pulleys, the large weight gains potential energy and the small weight loses potential energy. If friction is negligible, the gain of potential energy of the large weight is equal to the loss of potential energy of the small weight. If friction is not negligible, the large weight gains less potential energy than the small weight loses. In other words, there is a discrepancy between the loss of potential energy of the small weight and the gain of potential energy of the large weight. This discrepancy was investigated by James Joule in the nineteenth century. He made careful measurements and showed that the discrepancy is exactly accounted for by thermal energy. The unit of energy and work is named after Joule because his investigations and others that followed led to the conclusion that energy is conserved in the changes investigated. All the available evidence indicated that energy is always conserved. The Principle of Conservation of Energy was thus established and remains so to this day, although future discoveries might disprove the principle.*

Insight

Remember, energy is the capacity to do work. Conservation of energy means that the capacity to do work is unchanged.

Energy and power

On p. 51 we saw that the power output of a machine is given by the load × the speed at which the machine moves the load. This link can be applied to any object or machine being driven at constant speed. If the force is in newtons and the speed in metres per second, the power is in watts. For example, suppose the engine of a vehicle moving at a constant speed of 25 m/s exerts a driving force of 400 N. The power output of the engine is therefore 400 N × 25 m/s = 10 000 watts. In energy terms, the vehicle moves a distance of 25 metres every second, so the work done by the engine each second is therefore 10 000 joules (= 400 N × 25 m). The power output in watts is therefore the work done by the engine per second. In other words, one watt of power is an energy transfer rate of one joule per second. This general statement applies to any energy transfer process.

Power is rate of transfer of energy.

The unit of power is the watt (W), which is equal to a rate of transfer of energy of 1 joule per second. A 3000 watt electric

heater emits heat at a rate of 3000 joules every second.
A 100 watt light bulb uses electrical energy at a rate of 100 joules per second. Every minute, a 100 W light bulb would use 6000 joules of electrical energy. A weightlifter who raises an 800 N weight through a height of 1 metre in 2 seconds has a power output of 400 watts, because the weight is supplied with 800 J of potential energy (= weight × height gain) in 2 seconds.

$$\text{power (in watts)} = \frac{\text{energy transferred (in joules)}}{\text{time taken (in seconds)}}$$

The above equation may be rearranged to give:

energy transferred (in joules) = power (in watts) × time taken (in seconds).

How powerful are you?

Time how long it takes you to walk up a flight of stairs that you normally use. If you cannot use stairs, time a friend who can walk up the stairs. Then measure the height from the bottom to the top stair. If necessary, measure the height of a single step and multiply by the number of steps.

Multiply your weight in newtons by the total height gain, to give your gain of potential energy. Then divide the potential energy gain by the time taken, to give your power output in watts. Note that your weight in newtons is your mass in kilograms × g. For the purpose of this activity, use $g = 10$ m/s^2.

Worked example
A person of mass 54 kg walks up a flight of 20 stairs in 6 seconds. The height of each stair is 0.15 m. Calculate the power output of this person.

SOLUTION

Total height gain = 20 × 0.15 m = 3.0 m
Weight = 54 kg × 10 m/s² = 540 N
Gain of potential energy = 540 N × 3.0 m = 1620 J
Power output = $\dfrac{1620 \text{ J}}{6 \text{ s}}$ = 270 W

QUESTIONS

g = 9.8 m/s²

1 Describe the energy changes that take place when:
   (a) a pendulum bob swings from one side to the other side
   (b) a ball is released above a hard floor, hits the floor and
       rebounds to a lesser height.

2 (a) Calculate the gain of potential energy when a 500 N weight is
       raised by a height of 4.0 m.
   (b) How much electrical energy is supplied to a 100 watt light
       bulb in 20 seconds?

3 (a) Calculate the electrical energy used by a 3000 watt electric
       kettle in exactly 5 minutes.
   (b) How long would a 100 watt light bulb need to be switched on
       to use the amount of energy calculated in (a)?

4 A child of mass 40 kg climbs 3.0 m up a vertical rope in
   25 seconds. Calculate:
   (a) the weight of the child
   (b) the gain of potential energy of the child
   (c) the power output of the child's arm muscles.

5 At a swimming pool, a person of mass 55 kg jumps off a high diving
   board into the pool 4.5 m below the board.
   (a) Calculate the loss of potential energy of the person due to this
       vertical descent of 4.5 m.

(b) Describe the energy changes of the person after leaving the diving board.

Efficiency and power

▶ *The* **power output** *of a machine is the energy it supplies each second to the object it drives.*
▶ *The* **power input** *to a machine is the energy supplied to the machine each second.*

Each second, a certain amount of energy is supplied to the machine and the machine delivers a certain amount of energy to whatever it is driving. The energy per second delivered by the machine cannot exceed the energy per second supplied to the machine, because energy cannot be created in the machine. If all the energy delivered by the machine each second is equal to the energy supplied to it each second, the machine is perfectly efficient. In practice, friction is usually present between the moving parts of a machine and so most machines are not perfectly efficient.

The **efficiency** of a machine is defined as

$$\frac{\text{the energy per second delivered by the machine}}{\text{the energy supplied to the machine per second}}$$

Because the energy per second delivered by the machine is its power output, and the energy per second supplied to it is its power input,

$$\text{efficiency} = \frac{\text{power output}}{\text{power input}}$$

The maximum value of efficiency of a machine is 1. This is when the power output is equal to the power input and no power is wasted in the machine.

Note that the percentage efficiency of a machine is its efficiency × 100%.

Worked example

A 500 W electric winch raises a weight of 200 N at a speed of 0.4 m/s. Calculate:

(a) the electrical energy supplied to the winch each second

(b) the gain of potential energy of the weight in 1 second

(c) the percentage efficiency of the winch.

$$g = 9.8 \text{ m/s}^2$$

SOLUTION

(a) The power input to the winch = 500 watts. Therefore, in 1 second, the winch is supplied with 500 J of electrical energy.

(b) The weight gains a height of 0.4 m in 1 second. Hence the gain of potential energy in 1 second = weight × height gain in 1 second = 200 N × 0.4 m = 80 J.

(c) Efficiency = $\dfrac{\text{energy delivered by machine each second}}{\text{energy supplied to the machine each second}}$

$\qquad = \dfrac{80 \text{ J}}{500 \text{ J}} = 0.16$

Percentage efficiency = 0.16 × 100% = 16%.

> **Insight**
> Friction between the moving parts of a machine causes a heating effect. Friction transforms some of the energy supplied to a machine into thermal energy, thus reducing the efficiency of the machine.

MORE ABOUT KINETIC ENERGY

Kinetic energy is the energy possessed by a moving object due to its motion. Consider an object of mass m which accelerates from rest at constant acceleration a for a certain time t.

- *The speed of the object after time t, v = acceleration × time = at.*
- *Hence the distance moved s = average speed × time = ½ vt since the average speed = ½ v.*
- *Therefore the work done on the object = force × distance moved = (mass × acceleration) × distance moved = mass = $m(v/t) \times \frac{1}{2}vt = \frac{1}{2}mv^2$*
- *Because all the work done goes to kinetic energy, it follows that the kinetic energy of the object at speed v is equal to $\frac{1}{2}mv^2$.*

$$\text{Kinetic energy} = \frac{1}{2}mv^2$$

Worked example

Calculate the kinetic energy of a vehicle of mass 800 kg moving at a speed of 30 m/s.

SOLUTION

$$\text{Kinetic energy} = \frac{1}{2}mv^2 = \frac{1}{2} \times 800 \times 30^2 = 360000 \text{ J}$$

MORE ABOUT POTENTIAL ENERGY

The weight of an object of mass $m = mg$. To raise the object, an equal and opposite force must be applied to it. Therefore, if the object is raised through a height H, its potential energy increases by an amount equal to its weight × the gain of height, which is equal to mgH. The same formula is used for loss of potential energy, with H as the loss of height.

$$\text{Change of potential energy} = mgH$$

Worked example

A ball of mass 0.5 kg is released from rest at a height of 3.0 m above a floor. Calculate:

(a) the loss of potential energy of the ball just before it hits the floor,

(b) its speed just before it hits the floor, assuming that its kinetic energy is equal to the loss of potential energy.

$$g = 9.8 \text{ m/s}^2$$

SOLUTION
(a) Loss of potential energy = mgH = 0.5 × 9.8 × 3.0 = 14.7 J

(b) Its kinetic energy just before hitting the floor = 14.7 J.

$$\therefore \tfrac{1}{2}\,mv^2 = 14.7, \text{ where } v \text{ is its speed just before impact.}$$
$$\text{Hence } \tfrac{1}{2} \times 0.5 \times v^2 = 14.7$$
$$v^2 = \frac{2 \times 14.7}{0.5} = 59$$
$$v = 7.7 \text{ m/s}$$

Insight

Theme-park rides are good places to study energy transformations. Prove for yourself that a rollercoaster train that descends through a height H of 50 m would reach a speed v of 32 m/s (= 70 mph), assuming all the potential energy (mgH) is transformed into kinetic energy ($\tfrac{1}{2}mv^2$). Find out more by visiting a theme park or its website.

QUESTIONS

$g = 9.8\,m/s^2$

6   A 500 watt electric winch raises a weight of 400 N through a height of 2.5 m in 10 seconds. Calculate:
     (a) the gain of potential energy of the weight
     (b) the electrical energy supplied to the winch
     (c) the percentage efficiency of the winch.

7   A lift and its passengers have a total mass of 2800 kg.
     The lift takes 40 seconds to ascend a vertical distance of 20 m.
     (a) Calculate the gain of potential energy per second of the lift.

(b) A 25 kW electric motor is used to raise the lift. Calculate the efficiency of the motor when it is used to make the lift and its passengers ascend.

8 An elevator in a factory raises boxes of weight 200 N through a height of 5.0 m. When the elevator is operating at normal speed, it transports three boxes every minute from the bottom to the top of the elevator. Calculate:
   (a) the gain of potential energy of a single box on the elevator
   (b) the power output of the elevator motor.

9 (a) Calculate the kinetic energy of:
       (i) a 60 kg athlete running at a speed of 10 m/s
       (ii) a 2000 kg truck moving at a speed of 30 m/s
       (iii) a ball of mass 0.2 kg moving at a speed of 20 m/s.

   (b) Calculate the change of potential energy of:
       (i) a 50 kg person who jumps off a wall of height 0.80 m
       (ii) a ball of mass 0.2 kg thrown 12 m vertically into the air
       (iii) a 500 N weight raised through a height of 2.5 m.

10 At a fairground 'water chute', a small boat and its passengers descend through a height of 3.0 m on the chute. The boat reaches a speed of 5 m/s at the end of the chute. The total mass of the boat and its passengers is 300 kg. Calculate:
   (a) the loss of potential energy of the boat and its passengers
   (b) the kinetic energy of the boat at the end of the chute.

10 THINGS TO REMEMBER

1 **Work done** = *force × distance moved in the direction of the force.*

2 **Energy** *is the capacity to do work.*

3 *The unit of work and of energy is the* **joule** *(J).*

4 **Power** *is the rate of transfer of energy.*

$$power\ (in\ watts) = \frac{energy\ transferred\ (in\ joules)}{time\ taken\ (in\ seconds)}$$

5 *The unit of power is the* **watt** *(W). 1 W = 1 J/s.*

6 **Kinetic energy** *is energy due to motion* = $\frac{1}{2}\ mv^2$

7 **Potential energy** *is the energy of an object due to its position. Change of potential energy = mgH*

8 **Elastic energy** *is the energy stored in an object when it is stretched or squeezed.*

9 *Efficiency* = $\dfrac{the\ energy\ per\ second\ delivered\ by\ the\ machine}{the\ energy\ supplied\ to\ the\ machine\ per\ second}$

= $\dfrac{output\ power}{input\ power}$

10 *Energy is always conserved in any change.*

5

Thermal physics

In this chapter you will learn:

- *what is meant by absolute zero and where the coldest place in the Universe is*
- *how to work out the energy needed to heat an object or to melt a solid or boil a liquid*
- *how to reduce your home heating bills and save energy and money.*

Machines at work and at home do many jobs that would otherwise be tiring or time-consuming. The production of iron and steel and the invention of the steam engine started the industrial revolution of the nineteenth century. In the twentieth century, the internal combustion engine and the jet engine took over from the steam engine, making travel to any place on Earth possible. In this chapter we will look at the measurement of temperature, heat transfer and some of the thermal properties of materials, in preparation for a deeper look in Chapter 6 at the principles behind energy transformations by heat engines.

Heat and temperature

In winter in Siberia, the temperature outdoors can fall below −40°C. In summer in the Sahara desert, the temperature can rise above 40°C. These temperatures are expressed on the Celsius scale, denoted by the symbol °C (degree C). This scale is defined in terms of two 'fixed points', which are:

▶ *ice point, the melting point of pure ice at atmospheric pressure, 0°C*
▶ *steam point, the boiling point of pure water at atmospheric pressure, 100°C.*

A fixed point is a standard 'degree of hotness' which can be reproduced as required. A thermometer calibrated in °C would read 0 when in pure melting ice at atmospheric pressure, and 100 when in steam at atmospheric pressure. Figure 5.1 shows a mercury-in-glass thermometer. The temperature of the mercury in the bulb is measured by reading where the end of the mercury column is against the scale. The thermometer is used by placing the bulb at the location where the temperature is to be measured, then reading the scale after allowing sufficient time for the reading to stop changing.

mercury in bulb

mercury column

-10 0 10 20 30 40 50 60 70 80 90 100 110 °C

scale ranges from -10°C to 110°C

Figure 5.1 A mercury-in-glass thermometer.

The mercury thermometer works because the mercury in the bulb expands when it becomes warmer and moves along the air column in the glass tube. Mercury is a very suitable liquid for use in a thermometer because it expands evenly with increasing temperature. Coloured alcohol is used as the liquid in thermometers for low temperature measurements, because mercury freezes below -39°C.

All thermometers give the same reading at ice point and at steam point because they are calibrated to read 0°C and 100°C, respectively, at these two fixed points. Because the expansion of a liquid with increasing temperature varies from one liquid to another, the **gas thermometer** is chosen as the standard for calibrating temperatures between the fixed points. This type of thermometer consists of a gas-filled sealed bulb connected to a pressure gauge. The pressure of a gas in a sealed bulb increases if the gas temperature is raised and decreases if the temperature is lowered. The ratio of the pressure at steam point to the pressure at ice point is always the same, regardless of the amount or type of gas used. This is why the gas thermometer is used as the standard thermometer. All other thermometers are calibrated against a gas thermometer below ice point, between ice point and steam point, and above steam point.

ABSOLUTE ZERO

An object placed in a refrigerator is made colder by the refrigerator because the refrigerator removes energy from the object. How cold can an object be? The lowest temperature possible is called **absolute zero**. No energy can be obtained from an object at this temperature. The pressure of a gas would be zero at absolute zero, provided the gas did not liquefy as it was cooled. By measuring the pressure of a fixed amount of gas at steam point, then at ice point, the temperature for zero pressure (i.e. absolute zero) can be estimated. This temperature is –273°C. No matter which gas or how much gas is used, the same value is always obtained for absolute zero.

Here is an example of how this value is obtained from the pressure readings at ice point and at steam point. The readings are in kilopascals (kPa) which are explained on p. 172.

Steam point pressure reading = 250 kPa.

Ice point pressure reading = 183 kPa.

Therefore, a pressure change of 67 kPa (= 250 – 183 kPa) is caused by a temperature change of 100°C (from ice point to steam point), which means that a pressure change of 1 kPa requires a temperature change of 1.49°C (= 100/67).

To reduce the pressure from 183 kPa to zero would therefore require the temperature to be lowered by 273°C (= 183 kPa × 1.49°C/kPa) from ice point. Hence absolute zero = −273°C. Precise measurements give −273.15°C for absolute zero.

The ideas about absolute zero were developed by Lord Kelvin in the nineteenth century. The absolute scale of temperature in kelvins (K) is defined from two fixed points, which are:

▶ *absolute zero at zero kelvins (= −273°C)*
▶ *the temperature at which ice, water and water vapour coexist (= 0°C).*

Hence absolute temperature in kelvins = temperature in °C + 273.

The coldest place in the Universe

Low-temperature research laboratories are the coldest places in the Universe. Scientists measure ultra-low temperatures in microkelvins (millionths of a kelvin). Methods to achieve such low temperatures involve making a suitably cold object work to reduce its energy. Low-temperature research has produced some astonishing discoveries, including superconductors and superfluids.

HEAT TRANSFER

Heat is energy transferred owing to a temperature difference. If two objects in contact with each other are at the same temperature, no heat transfer between the two objects takes place. The two objects are said to be in thermal equilibrium, because there is no heat transfer between them. 'The baby in the bath' rule is based on the idea of thermal equilibrium; the parent holds the baby and dips his or her elbow (not the baby's elbow!) in the bathwater to test the water temperature. If the water temperature is the same as the elbow temperature, the baby can then be dipped in the water because he or she is in thermal equilibrium with the parent.

Heat transfer can occur in three different ways:

1   **Thermal conduction** *takes place in solids, liquids and gases when there is a temperature difference between different parts of the substance. Metals are the best conductors of heat for the same reason that they are good conductors of electricity; they contain tiny particles called electrons, which can move about carrying energy from one part of the metal to another part when there is a voltage or a temperature difference across the metal. We will meet the electron many times in this book.*
    Insulating materials such as polystyrene sheets are made of substances in which all the electrons are trapped inside the atoms of the substance and cannot move through the substance. The presence of air pockets in an insulating material improves its insulation properties.
2   **Thermal convection** *is the process of circulation that takes place in liquids and gases when there is a temperature difference between different parts of the liquid or gas. The density of a liquid or gas varies with temperature, so temperature differences in a liquid or gas cause density variations. Gravity makes denser regions sink and less dense regions rise, thus causing circulation currents. A hot air balloon rises because the air in the balloon is heated by a burner and so becomes less dense and is therefore forced upwards. Vehicle cooling systems transfer heat from the engine to a radiator as a result of circulation of water heated by the engine and cooled in the radiator. The radiator is designed to lose heat to the surroundings by convection of air, assisted by a fan if necessary.*
3   **Thermal radiation** *is emitted from every surface. The hotter the surface is, the greater the thermal radiation from it.*

No *substance is needed to carry the radiation as it can travel through a vacuum.*

> A *black surface is a much more effective absorber of thermal radiation than a shiny silvered surface. You can notice this if you hold a piece of black paper and a sheet of tinfoil at the same distance from a hot object. The black paper warms up much faster than the tinfoil because it absorbs radiation from the hot object much more effectively.*

> A *black surface is a much more effective radiator of thermal radiation than a shiny silvered surface. If you want to keep an object warm, wrap it in tinfoil so that it does not lose energy due to thermal radiation. If it is wrapped in black paper, it will cool down very quickly. A vehicle radiator is usually painted black to enable it to lose heat by radiation as well as by convection.*

Insight

Shiny silvered blankets are used by rescue teams in cold weather to keep people warm. The outer silvery surface radiates much less thermal radiation than any other surface would.

HOW TO REDUCE YOUR FUEL BILLS

Heat flow is measured in watts, where 1 watt is equal to a rate of energy transfer of 1 joule per second. A 1000 watt (= 1 kilowatt) heater supplies heat to its surroundings at a rate of 1000 joules per second. Fuel usage on fuel bills is usually priced in **kilowatt hours** (kWh) regardless of what the fuel is used for (e.g. heating, lighting, running a computer). One kilowatt hour is the energy used by a 1 kilowatt electric heater in 1 hour. For example, a gas bill that shows the fuel used as 11 500 kWh tells you that the amount of gas used gave the same energy as a 1 kilowatt heater for 11 500 hours. A fuel bill usually states the price per kilowatt hour of the fuel. For example, the cost of 11 500 kWh of fuel at a price of 3.0 p per kWh would be £345. Look at a recent fuel bill and work out how much fuel you use in your home each day and what the cost is.

Home heating bills can be reduced by fitting loft insulation and double-glazing to reduce heat loss from indoors to outdoors. The picture of the house below shows these and other measures that can be taken to reduce fuel bills.

Figure 5.2 Reducing home heating bills.

To work out if an installation such as double-glazing is cost-effective, the cost of the installation needs to be compared with an estimate of the amount that would be saved on fuel bills. Such an estimate can be made if the **U-value** of the installation is known. This is the heat flow (i.e. heat transfer per second) per square metre for a temperature difference of 1°C.

▶ *Loft insulation has a U-value of about 0.5 watts per square metre per °C. This means that the heat flow through 1 square metre of loft insulation is 0.5 watts for a temperature difference of 1°C. The heat flow through 50 square metres of this loft insulation would be 250 watts for a temperature difference of 10°C. The U-value of a 50 m² roof without loft insulation is about four times greater; thus a temperature difference of 10°C would cause 1000 watts of heat loss. Fitting loft insulation would reduce the heat loss by 750 watts or 0.75 kilowatts and so, in 90 days, the saving would amount*

*to over 1600 kWh (= 0.75 kW × 24 hours/day × 90 days).
At 3.0 p per kWh, the householder with loft insulation would
therefore save about £48. Clearly, the calculation assumes an
average temperature difference of 10°C.*

▶ *Double-glazing reduces heat loss due to conduction because
a layer of air is trapped between two glass panes. Air is a
much better insulator than glass and therefore a double-
glazed window cuts heat loss by a factor that depends on
the thickness of the layer of trapped air. For a temperature
difference of 10°C, a typical single-pane window loses heat at
a rate of about 50 watts per square metre, whereas a double-
glazed window loses heat at a much smaller rate of about
30 watts per square metre.*

▶ *Cavity wall insulation consists of plastic foam which is
pumped into the cavity between the two brick walls and
then sets. This provides a very effective insulating layer
between the walls. A typical cavity wall without insulation
in the cavity loses heat at a rate of about 15 watts per square
metre for a 10°C temperature difference; cavity wall insulation
can reduce this by a factor of three to about 5 watts per square
metre.*

▶ *Other less expensive measures to reduce heating bills include:*

   1 *draught-proofing doors to cut down heat loss due to
convection*

   2 *placing aluminium foil behind any radiator attached to
an outside wall to cut down thermal radiation from the
radiator to the wall*

   3 *closing the curtains (at night!) to cut down heat loss due
to radiation through the window.*

Insight

Better home insulation reduces your **carbon footprint** because
you use less energy to heat your home in cold weather. Visit
a carbon footprint calculator website to see how your carbon
footprint would be reduced (or has been reduced) by better
home insulation.

QUESTIONS

1 State the temperature of (a) ice point, (b) steam point, (c) absolute zero on:
(i) the celsius scale, (ii) the absolute scale.

2 Explain why (a) a saucepan has a plastic or wooden handle, (b) the outlet pipe from a hot water tank is connected to the top of the tank, (c) a hot potato in tinfoil stays hot much longer than an unwrapped hot potato.

3 (a) A semi-detached house (Number 1) has an uninsulated roof of area 60 m^2, single-pane windows of total area 10 m^2 and external walls without cavity insulation of total area 120 m^2. Copy the table below, then estimate and fill in 'Number 1's rate of heat loss' through the roof, windows and walls for a 10°C temperature difference.

Temperature difference = 10°C	U-value $Wm^{-2}°C^{-1}$	Area m^2	Number 1's rate of heat loss W	Number 3's rate of heat loss W
Uninsulated roof	2.0	60		—
Insulated roof	0.5	60	—	
Single-pane window	5.0	10		—
Double-glazed window	3.0	10	—	
External wall with no cavity wall insulation	1.5	120		—
External wall with cavity wall insulation	0.5	120	—	
Total rate of heat loss	—	—		
Total cost per day at 3.0 p/kWh	—	—		

(b) The adjoining semi-detached house (Number 3) has loft
insulation, double-glazing and cavity wall insulation. Estimate
and fill in 'Number 3's rate of heat loss' for a 10°C temperature
difference.
(c) Both houses are heated by gas-fired central heating at a cost of
3.0 p per kWh. Estimate the difference between the cost of gas
in one day for these two houses.

Thermal properties of materials

Metals heat up much more easily than most non-metals, which
is why the interior of a parked vehicle can become very hot in
summer. The energy possessed by a substance due to its temperature
is referred to as **thermal energy**. All substances are composed of
molecules, which are the smallest particles of a substance. In solids,
the molecules are held together rigidly by strong **bonds**. In liquids,
the bonds are not strong enough to keep the molecules in a rigid
structure. In a gas or a vapour, the molecules move about freely
at large distances from each other. See pp. 175–6 for more
information. When energy is supplied to a substance to increase
its thermal energy, the particles of the substance

▶ *gain kinetic energy, if the temperature of the substance increases*
▶ *use the energy supplied to break the bonds between the
molecules, if the substance changes state from a solid to
a liquid or to a gas, or from a liquid to a gas.*

MELTING POINTS AND BOILING POINTS

When a pure solid is heated and heated, its temperature increases
until it reaches the **melting point** of the solid. The solid changes
to a liquid at this temperature, provided it continues to be heated.
Continued heating raises the temperature of the liquid to its **boiling
point,** the temperature at which the liquid boils. The melting point
of a solid is characteristic of the solid and the boiling point of a
liquid is characteristic of the liquid and therefore can be used to

identify the substance. Freezing is the reverse process to melting and it occurs when a liquid has been cooled to its freezing point, which is the same temperature as the melting point. For example, the freezing point of water is 0°C, the same as the melting point of pure ice. The reverse process to boiling is condensation. If a gas in a container is cooled sufficiently, liquid forms on the surface of the container due to condensation.

Winter journeys

Roads in winter are usually covered with grit and salt if freezing weather is due. Icy roads are dangerous because vehicles are difficult to control on ice due to lack of grip on the road surface. Salt dissolves in water and salty water has a lower freezing point than pure water. If the freeze is severe, grit particles in the ice help to provide some grip. Nevertheless, vehicle speeds must be greatly reduced in freezing weather conditions to reduce the risk of traffic accidents.

SPECIFIC HEAT CAPACITY

The **specific heat capacity** of a material is the energy needed to raise the temperature of unit mass of the material by one degree. For example, the specific heat capacity of water is 4200 joules per kilogram per °C. This means that:

▶ *4200 J of energy must be supplied to raise the temperature of 1 kilogram of water by 1°C*
▶ *42 000 J of energy must be supplied to raise the temperature of 1 kilogram of water by 10°C*
▶ *84 000 J of energy must be supplied to raise the temperature of 2 kilograms of water by 10°C*

More generally, to raise the temperature of mass m of a substance from T_1 to T_2,

> *the energy needed* $\Delta E = mc\,(T_2 - T_1)$,

where c is the specific heat capacity of the material. The unit of c is J/kg/°C.

Notes

1 *To calculate c using the formula $\Delta E = mc(T_2 - T_1)$, rearrange the formula to make c the subject as follows:*

$$mc\,(T_2 - T_1) = \Delta E$$
$$\text{Therefore } c = \frac{\Delta E}{mc\,(T_2 - T_1)}$$

2 *Specific heat capacities of some other substances are listed in the following table.*

Substance	Specific heat capacity in J/kg/°C
Water	4200
Copper	380
Aluminium	900
Ice	2100

Insight

Water has a very high specific heat capacity compared with other substances. This is why sea temperatures change little compared with ground temperatures from summer to winter.

How to measure the specific heat capacity of a metal

You will need a thermometer, a 12 V power supply, a joulemeter, a 12 V electric heater, an insulated metal block with slots for the heater and the thermometer, and a balance.

Figure 5.3 Measuring the specific heat capacity of a metal.

1 *Use the balance to measure the mass m of the metal block to be tested and write down your measurement.*
2 *Connect the heater to the joulemeter and connect the joulemeter to the 12 V power supply, but do not switch on the power supply yet. Place the heater and the thermometer in the slots in the metal block. Measure and note the initial temperature T_1 of the block and the initial reading of the joulemeter.*
3 *Switch on the power supply for five minutes. Note the joulemeter reading at the end of this time and the temperature T_2 of the block.*
4 *Calculate the energy supplied, ΔE, in joules, from the difference of the joulemeter readings and calculate the temperature difference $(T_2 - T_1)$.*
5 *Use the formula $\Delta E = mc(T_2 - T_1)$ to calculate the specific heat capacity c of the block.*

Sample measurements for an aluminium block:

Mass of block m = 1.0 kg

Initial temperature T_1 = 15°C

Highest temperature T_2 = 42°C

Initial reading of the joulemeter = 22 500 J

Final reading of the joulemeter = 46 800 J

Show that the results give the specific heat capacity of aluminium = 900 J/kg/°C.

> To raise the temperature of mass m of a substance from T_1 to T_2, the energy needed $\Delta E = mc(T_2 - T_1)$, where c is the specific heat capacity of the material. The unit of c is J/kg/°C.

QUESTIONS

Use the specific heat capacity values given in the table on p. 76 in the following questions.

4 (a) Calculate the energy needed to heat 1.5 kg of water from 10°C to 100°C.
   (b) Calculate the energy needed to heat an aluminium kettle of mass 0.4 kg from 10°C to 100°C.
   (c) Calculate the energy needed to heat 1.5 kg of water in an aluminium kettle of mass 0.4 kg from 10°C to 100°C.

5 (a) Calculate the final temperature when 2500 J of thermal energy is supplied to a copper block of mass 0.51 kg at an initial temperature of 12°C.
   (b) An insulated aluminium block of mass 2.0 kg is heated by a low-voltage 36 W electric heater for 10 minutes. Calculate:
       (i) the energy supplied by the heater
       (ii) the temperature rise of the aluminum block.

6 An insulated copper hot water tank has a mass of 20 kg and contains 25 kg of water at 15°C.
   (a) Calculate the energy needed to heat the tank and the water to 45°C.
   (b) Calculate how long a 3.0 kW electric immersion heater in the tank would take to supply the energy calculated in (a).

SPECIFIC LATENT HEAT

Energy must be supplied to a solid to melt it. There is no temperature rise at the melting point because the energy supplied is used by the molecules in the solid state to break the bonds between them. Energy must be supplied to a liquid at its boiling point to boil the liquid away. The energy supplied is used by the molecules in the liquid state to break the bonds between them. Because there is no temperature rise at the melting point or the boiling point, the energy supplied is called **latent heat**. The word 'latent' means 'hidden'.

The **specific latent heat** of a solid or a liquid substance for a given change of state is the energy needed to change the state of unit mass of material, without change of temperature. For example, the specific latent heat of fusion of ice is 336 000 joules per kilogram. This means that:

▶ *336 000 joules of energy is needed to melt 1 kilogram of ice*
▶ *672 000 joules of energy is needed to melt 2 kilograms of ice*
▶ *1.68 million joules of energy is needed to melt 5 kilograms of ice.*

More generally, to change the state of mass m of a substance at constant temperature,

$$\text{the energy needed } \Delta E = ml$$

where l is the specific latent heat of fusion (for melting or solidifying) or vaporization (for vaporizing, boiling or condensing) or sublimation (for a solid which vaporizes directly or a vapour which forms a solid without liquefying first) for that substance. The unit of l is J/kg.

Note: To calculate l using the formula $\Delta E = ml$, rearranging the formula to make l the subject gives $l = \Delta E/m$.

How to measure the specific latent heat of steam

Use an electric kettle with a known power rating in watts. Fill the
kettle with a measured volume of water. Switch on the kettle and
boil the water in it for 2 minutes. You will need to open some
windows when you do this to prevent condensation in the room.
After the 2 minutes, switch off the kettle and let the water cool.
When it is sufficiently cool, measure the volume of the water that is
still in the kettle. The difference between the volume of water at the
start and the volume at the end is the volume of water boiled away.
Calculate the mass of water boiled away, given that the density of
water is 1.0 kilogram per litre.

Calculate the energy supplied to the kettle from the power rating in
watts × 120 seconds.

Determine the latent heat of steam from the energy supplied and
the mass of water boiled away.

Sample measurements

Initial volume of water = 1.50 litres

Final volume of water = 1.35 litres

Power of kettle = 3000 watts

Energy supplied in 2 minutes = 3000 watts × 120 seconds =
360 000 joules

Volume of water boiled away = 0.15 litres

Mass of water boiled away = density of water × volume boiled
away = 1.0 kg/litre × 0.15 litres = 0.15 kilograms

Therefore, the specific latent heat of steam = 360 000 joules/ 0.15 kilograms = 2 400 000 joules per kilogram.

QUESTIONS

The specific latent heat of ice = 336 000 J/kg.

The specific latent heat of steam = 2.3 MJ/kg.

The specific heat capacity of water = 4200 J/kg/°C.

7 (a) Calculate the energy needed to boil away 0.5 kg of water.
   (b) Calculate the mass of ice that could be melted, given the amount of energy calculated in (a).

8 On a cold sunny day, a bucket containing 2.5 kg of ice at 0°C melts and warms to a temperature of 5°C.
   (a) Calculate the energy needed to (i) melt 2.5 kg of ice at 0°C, (ii) heat the melted ice to 5°C.
   (b) The bucket of ice takes 3.5 hours to melt and warm to 5°C. Calculate the energy per second gained by the bucket.

9 A plastic beaker containing 0.20 kg of water at 15°C placed in a freezer cools down and freezes within 20 minutes. Calculate:
   (a) the energy removed from the water to (i) cool it down to 0°C, (ii) freeze it
   (b) the rate at which energy is removed from the water.

10 A 3000 watt electric kettle is used to boil water. Calculate:
   (a) the electrical energy supplied to the kettle in 1 minute
   (b) the mass of water boiled away in 1 minute. Assume all the energy supplied is used to boil the water.

10 THINGS TO REMEMBER

1 *The* **melting point** *of pure ice is* 0°C *and the* **boiling point** *of water at atmospheric pressure is* 100°C.

2 **Absolute zero,** *the lowest possible temperature, is* –273°C. **Absolute temperature** *in kelvins = temperature in* °C + 273.

3 *Heat transfer occurs owing to thermal* **conduction, convection** *and* **radiation.**

4 *Metals are much better heat conductors than non-metals.*

5 *When part of a liquid or a gas is heated, the substance becomes less dense and rises, causing thermal convection.*

6 *Thermal radiation is emitted from the surface of an object. The hotter the surface, the greater the intensity of the radiation it emits.*

7 *A matt black surface is a better absorber and emitter of thermal radiation than a shiny silvery surface.*

8 *Home insulation measures include loft insulation, cavity wall insulation and double-glazing.*

9 *The* **specific heat capacity** *of a substance is the energy needed to raise the temperature of 1 kg of the substance by* 1°C.

10 *The* **specific latent heat** *for a given change of state of a substance is the energy needed to melt or boil away 1 kg of the substance.*

6

Engines and thermodynamics

In this chapter you will learn:
- *how heat engines work, including the steam engine, the car engine and the jet engine*
- *about the laws of thermodynamics, the rules that govern energy transfers*
- *why energy needs to be used more efficiently than at present*
- *what energy resources will be available in the future.*

The science of thermodynamics is about temperature, heat and work and it provides rules and limits on the use of energy. At the start of the nineteenth century many scientists thought of heat as an invisible fluid, referred to as 'caloric'. Two decades later the caloric theory had been consigned to oblivion, replaced by the new theory that heat is a form of energy. By the middle of the nineteenth century scientists had worked out the laws of thermodynamics, including the all-important second law, which tells us that energy tends to spread out and become less useful as a result. In this chapter we will look at these ideas and how they govern the efficiency limits of engines. In addition, we will make a survey of present and future energy supply and demand, including fuel supplies and renewable energy resources.

Heat engines

Before the invention of the steam engine, power to drive machinery was provided by windmills or watermills and power for transport was provided by horses on land and wind and ocean currents at sea. James Watt recognized the potential of steam power, as a result of observing jets of steam from a boiling kettle. The invention and development of the steam engine created industries that were able to produce and distribute goods and food on a much larger scale than had ever been possible previously. Nowadays, the steam engine has faded into obscurity, replaced by engines such as the internal combustion engine, the diesel engine, the electric motor and the jet engine. Steam or gas turbines drive electricity generators in power stations to supply electricity to homes, industry and rail transport. Although an electric motor is not an engine, as it does not have its own source of energy, the electricity for it is supplied by an electricity generator driven by a turbine in a power station. Thus we rely on engines of one form or another for transport and for our energy needs at home and at work. Most types of engines are heat engines, as they obtain energy by burning fuel. An exception is a turbine in a hydroelectric power station, which is driven by water flowing downhill from an upland reservoir. The aerogenerator, a wind-driven electricity generator, is another exception.

Steam engines and turbines, internal combustion engines, jet engines and rocket engines are described as **heat engines** because each type of engine uses energy from a high-temperature source to do work. In other words, a heat engine uses heat to do work.

To drive a **steam engine**, burning fuel is used to boil water in a boiler, so producing steam at high pressure. The steam is fed via pipes and valves to cylinders, each containing a piston coupled to a drive shaft. The valves are opened and closed automatically by rods connected to the drive shaft. Thus the pressure of the steam drives the piston back and forth along the cylinder, rotating the drive shaft in the process.

In a four-stroke internal combustion engine, an electric spark is used to explode a mixture of gasoline and air, forcing the piston to move along the cylinder and turn the drive shaft. An inlet and an outlet valve are opened and closed automatically in a sequence of four piston strokes, as follows:

1 **Squeeze stroke:** *inlet valve closed; outlet valve closed; the piston moves into the cylinder and compresses the fuel mixture in the cylinder. A spark explodes the mixture at maximum compression, causing the temperature and pressure of the air in the cylinder to shoot up.*
2 **Power stroke:** *inlet valve closed; outlet valve closed; the piston is forced out by the high pressure in the cylinder. The pressure in the cylinder drops as the piston moves out.*
3 **Sweep stroke:** *inlet valve closed; outlet valve opens; the piston returns and forces the combustion products out of the cylinder via the outlet valve.*
4 **Draw stroke:** *inlet valve open; outlet valve closed; the piston moves out of the cylinder (driven by the power stroke of one of the other cylinders) and draws more fuel and air into the cylinder for the next 'squeeze' stroke.*

Figure 6.1 The internal combustion engine.

In a **steam turbine**, steam jets are directed at the blades of a turbine wheel, causing the wheel to rotate. The wheel is usually one of several wheels on the same shaft inside an enclosed turbine block. The base of the block is kept cool by means of water pumped through pipes in the block.

In a **jet engine,** air is drawn into the engine and fuel from the feed pipes enters the air stream. The fuel mixture is burned downstream continuously in the burner chamber, thus raising the air pressure and temperature enormously. Hot air and combustion products emerge at high speed through the outlet duct into the atmosphere, thus forcing the engine forward.

Insight

A jet engine exerts a force on the hot gases to push them out. The hot gases exert an equal and opposite force on the engine which pushes the engine forwards.

Figure 6.2 The jet engine.

Heat transfer requires a low-temperature 'sink' as well as a high-temperature source. In the examples above, the atmosphere provides the low-temperature sink, either directly or indirectly, via a cooling system. Not all the heat from the high-temperature source can be used by the engine to do work. This restriction is because the low-temperature sink must accept some of the energy transferred from the high-temperature source, otherwise the engine would overheat and stop working. For example, in a steam turbine, steam jets at high pressure are used to drive the turbine wheel; the turbine is kept cool by means of water pumped through pipes that pass through the engine block. If this cooling system was switched off, the engine temperature would rise. The pressure inside the engine would therefore rise and this would lessen and eventually cancel the force of the steam jets that drive the turbine wheels.

THE EFFICIENCY OF A HEAT ENGINE

A heat engine does work as a result of using heat from the combustion of fuel. Some of the heat from the fuel is not used to do work and is supplied to the low-temperature 'sink'. The energy used to do work is referred to as 'useful energy'. The energy supplied to the low-temperature 'sink' is wasted by the engine as it is not used to do work. The efficiency of a heat engine is the fraction of the energy supplied to it from the fuel that is used to do work. For example, an engine with an efficiency of

▶ *0.2 converts 20% of the energy from its fuel into useful energy and wastes 80% of the energy from the fuel*
▶ *0.5 converts half the energy from its fuel into useful energy and the other half is wasted*
▶ *0.8 converts 80% of the energy from its fuel into useful energy and wastes 20% of the energy from the fuel.*

The principles of heat engines were established in the early nineteenth century by Sadi Carnot, a French scientist. Carnot realized that the lower the temperature of the 'sink', the more efficient the engine would be. This idea is like a hydroelectric scheme where water from an upland lake flows through pipes into a generator and then into the sea. The lower the generator, the more work the water does as it passes through the generator. The overall efficiency depends on the height drop from the upland lake to the generator. However, just as there is a limit to the drop between the upland lake and the generator, there is a limit to how cold the heat sink could be, namely the lowest possible temperature, which is absolute zero. Carnot worked out that for any heat engine operating at maximum efficiency between a high-temperature source and a low-temperature sink:

$$\frac{\text{the heat supplied to the sink}}{\text{the heat supplied by the source}} = \frac{\text{absolute temperature of the sink}}{\text{absolute temperature of the source}}$$

For example, if the engine was operating between 1000 K and 300 K,

$$\frac{\text{the heat supplied to the sink}}{\text{the heat supplied by the source}} = \frac{300}{1000} = 0.30$$

In this example, for every 100 joules supplied from the source, 30 joules would be taken by the sink. The work done by the engine in this situation would therefore be 70 joules from every 100 joules supplied by the source. In other words, the efficiency of this heat engine is 0.7.

Carnot proved that, for any heat engine operating between a source at temperature T_1 and a sink at lower temperature T_2,

$$\text{the maximum efficiency of such an engine} = \frac{(T_1 - T_2)}{T_1}$$

Thus an internal combustion engine in which the air is heated to 700 K (when the fuel mixture is ignited) and cooled to 300 K has a maximum efficiency of 0.57 (= (700 − 300)/700).

If combustion of fuel in this engine releases energy at a rate of 50 kilowatts (= 50 000 joules per second), the maximum power output of the engine would be 28.5 kW (= 50 kW × 0.57) and the energy per second wasted would be 21.5 kW (= 50 kW − 28.5 kW).

Insight

Friction between the moving parts reduces the efficiency of a heat engine further.

The laws of thermodynamics

Carnot's ideas about heat engines were developed into the laws of thermodynamics, which are the principles behind energy transformations involving heat.

1 *The* **First Law of Thermodynamics** *states that when an object or a system of objects does work or gains heat, the change of* **internal energy** *of the object(s) is equal to the difference between the heat gained and the work done by the object(s).*

   ▷ **Work** *is energy transferred by means of a force when the force moves its point of application in the direction of the force. See p. 52.*

   ▷ **Heat** *is energy transferred due to a difference of temperature.*

   ▷ *The* **internal energy** *of an object is the energy it possesses, not counting its kinetic or potential or elastic energy. For example, if the temperature of a solid object is increased by heating the object, the internal energy of the object increases as its atoms vibrate more.*

   The First Law of Thermodynamics follows from the Principle of Conservation of Energy, as heat and work are the means by which an object gains or loses energy. If an engine does work W as a result of accepting heat Q_1 from a high-temperature source and rejecting heat Q_2 to a low-temperature sink, then $W = Q_1 - Q_2$.

2 *The* **Second Law of Thermodynamics** *states that energy tends to spread out and become less useful whenever it is transferred between objects in an isolated system.*

   For example, when an electric winch is used to raise a weight, energy is transferred from the electrical supply to the weight. However, some of the energy supplied is wasted due to friction between the moving parts and due to the heating effect of the electric current in the circuit. This wasted energy cannot be recovered and used to do work as it spreads out by means of heat transfer to the surroundings. The energy from the electrical supply that is gained by the weight is an example of useful energy. The energy from the electrical supply that is not gained by the weight is wasted as heat transfer to the surroundings.

For a heat engine operating at its maximum efficiency, the work done by the engine must be less than the heat supplied from the high-temperature source. As we saw on p. 86, this is because heat transfer from the high-temperature source would stop if heat transfer to the low-temperature sink was prevented. Some heat transfer to the low-temperature sink must take place and therefore some energy from the source is wasted and not used to do work.

Insight

In effect, the First Law of Thermodynamics tells us that energy is always conserved and the Second Law tells us that energy tends to spread out in any change.

ENERGY FOR THE FUTURE

The problem with energy is that it tends to spread out. Even when energy is stored, as happens when a car battery is charged or when a weight is raised, some of the energy used in the storage process is wasted owing to friction or resistance and cannot be recovered. Although energy cannot be created or destroyed, it tends to spread out when it is used. Because our energy resources are finite, such resources will eventually run out. Each person in Europe and North America uses energy at an average rate of about 8000 joules per second, which amounts to 250 000 million joules per person each year. For a world population of about 6000 million people (likely to exceed 8000 million by 2050), each using energy at a rate of 8000 joules per second, the rate of energy usage would therefore be about 1500 million million million joules each year. At present, world energy use is about 300 million million million joules per year or about 1500 joules per second per person. Global energy use will probably double by 2050 as there will be more people on the planet and each person will use more energy as living standards rise.

Fossil fuels

Fossil fuels such as coal, oil and gas provide most of the energy we use at present. These fossil fuels are substances which release energy when burned in air. Fossil fuels are extracted from the Earth, having taken millions of years to form as a result of geological changes which have buried and compressed dead vegetation to form coal and dead marine life to form oil and gas. The energy released when fossil fuel is burned was stored millions of years ago by a living object which took in light from the Sun. Fuel is not renewable; once burned, the energy is released and the substance is changed into other compounds such as carbon dioxide and carbon monoxide. Each kilogram of fossil fuel releases about 30 million joules when burned. Approximately 10 million million kilograms of fossil fuel are burned each year to meet worldwide energy demands. The consequences of burning so much fossil fuel may be catastrophic.

▶ **Global warming** *is taking place because gases such as carbon dioxide are so-called* '**greenhouse gases**', *which trap heat radiation in the atmosphere. Many scientists reckon that if we don't reduce our carbon emissions into the atmosphere, global warming will cause devastating climate changes. Some scientists think it may already be too late and we should prepare for the worst.*

▶ *The world reserves of fossil fuels are reckoned to be about 1500 million million kilograms, sufficient to last for about 150 years. However, oil and gas, which constitute about 30% of present fossil fuel reserves, are being used at a faster rate than coal and will probably be used up within the 21st century. If industrial countries do not develop non-fossil energy resources soon, we could face a* **low-energy future** *as well as catastrophic* **climate changes**.

Nuclear reactors

Nuclear reactors release energy from uranium contained in fuel rods in the reactor. The energy released is removed by a coolant which is pumped through sealed pipes passing through the reactor core. The hot coolant flows through a heat exchanger, where it is used to raise steam, before being pumped back cooler nto the reactor core. The steam from the heat exchanger is used to drive turbines to produce electricity. A large nuclear reactor produces about 1500 megawatts of electrical power. See p. 218 for more details. Note that 1 megawatt (MW) = 1 million watts.

See p. 218 for more details.

Insight

A 1500 MW power station can supply electricity to about 200 000 homes.

At present, about 8% of worldwide energy demand is met by nuclear power stations, as they provide between a quarter and a third of the electricity supplies in industrial countries. World reserves of uranium will probably last about 50 years at the current rate of use. However, nuclear reactors at present use a form of uranium known as uranium 235, which is no more than about 2% of natural uranium. The other 98% of natural uranium is uranium 238, which is not used in the present type of nuclear reactors. The lifetime of the world's reserves of uranium would be extended many times if so-called 'fast breeder' reactors were to be built, making use of plutonium made in the present generation of nuclear reactors from uranium 238.

The used fuel rods from nuclear reactors are highly radioactive and remain so for thousands of years. Radioactivity is harmful to health as it is known to cause cancer. Therefore, safe and secure storage of radioactive waste from spent fuel rods is essential. The first and second generation of nuclear reactors currently in use will need to be replaced when they reach the end of their designated working life; otherwise, electricity cutbacks will probably happen – unless alternative means of generating electricity on a large scale are developed.

Fuel use and fuel reserves

World fuel use is about 500×10^{18} J per year. This is presently met from energy sources as shown in the following table. The table also shows estimates of how many years present fuel reserves will last, at the year 2007 rate of use.

World fuel use and reserves, 2007

Oil	Coal	Natural gas	Nuclear power	Hydroelectricity and other renewable sources
35%	29%	24%	6%	6%
about 50 years	<200 years	50–100 years	60 years*	Indefinitely

*for thermal reactors (3000 years for fast breeder reactors)

The total energy used by all sources in the **United Kingdom** is about 3% of the world fuel use, or about 15×10^{18} J per year.

UK fuel use in 2007 as a percentage of total UK energy use in 2007

Oil	Coal	Natural gas	Nuclear power	Hydroelectricity and other renewable sources
36%	17%	39%	6%	<2%

Imports of oil at present are about the same as exports. Net imports provide about 75% of the coal used and about 20% of the gas used. UK reserves of oil and gas are likely to last less than 20 years. Coal reserves would last considerably longer if coalfields closed since 1983 were re-opened.

Renewable energy resources
Renewable energy resources such as hydroelectric power stations and wind farms of wind turbines will need to be developed on a massive scale to meet our electricity demands when oil and gas reserves are used up, if more nuclear reactors are not constructed. However, renewable resources in general do have adverse environmental effects, as they need to cover a very large area to generate the same power as a fossil fuel or nuclear power station.

A renewable energy resource is a source of useful energy that does not require fuel. Most renewable energy resources use solar energy, either directly, as in solar panels, or indirectly, via the Earth's atmosphere which is heated by the Sun. Wind turbines are driven by winds created in the atmosphere, wave generators are driven by waves created by winds, hydroelectric generators are driven by rainwater running downhill. Other renewable resources include geothermal energy, which is obtained by pumping water into hot underground rock basins and using the steam generated to drive turbines. Tidal power is another non-solar energy resource. The tides rise and fall twice each day because the Moon pulls on the Earth's oceans as the Earth spins. A tidal power station consists of a barrier which traps seawater at high tide and releases it through channels in the barrier to drive turbines and produce electricity.

Could **solar energy** be used, directly or indirectly, to meet our energy needs? On a clear day, each square metre of the Earth's surface receives 1400 joules every second from the Sun, when the Sun is directly overhead. Taking account of cloud conditions and latitude, a solar panel fitted to a house roof in the northern hemisphere receives, on average, about 100 joules per second per square metre in the daytime. Each person would need about 80 square metres of solar panels to meet his or her energy needs. Solar panels could make a contribution to our energy needs, but would need to cover an enormous area to have a significant effect. A solar heating panel fitted to a house roof uses sunlight to heat water that flows at a steady rate through the panel. The hot water collects in an insulated tank connected to the hot water taps in the house. Solar cell panels consist of solar cells that generate electricity directly from sunlight. Fuel for vehicles might in future be produced on a large scale by using solar cells to make hydrogen gas from water, a process that involves no more than passing electricity through water. Vast arrays of solar cells covering large areas would still be necessary though.

Wind turbines are electricity generators on top of tall towers, each generator turned by propellers driven by the wind. A large wind turbine is capable of producing about a megawatt (= 1 million watts) of electrical power. A wind farm would need thousands of wind turbines to produce the same amount of power as a 5000 megawatt nuclear power station.

Hydroelectric power stations in hilly areas where there is plenty of rainfall contribute to the electricity supplies in some countries, mostly on a small scale at present. A single hydroelectric power station is not likely to produce much more than about 100 megawatts, but a large number of such stations could make a significant contribution to our energy needs.

A **tidal power station** in a suitable coastal location could provide as much electricity as a large power station but it would need to cover a large area. A suitable tidal area of 400 square kilometres could trap 4000 million cubic metres of water each high tide with a height difference of 10 metres between high tide and low tide. The density of water is 1000 kilograms per cubic metre, so the mass of such a large volume of water is 4 million million kilograms. As explained on p. 61, this amount of water would release 80 million million joules of potential energy for a 2-metre drop of height. Over the 12 hours between successive high tides, the average rate of release of energy would therefore be about 2000 million joules per second, sufficient to generate 1000 megawatts of electrical power from electricity generators operating at 50% efficiency.

Wave generators floating offshore could also make a significant contribution to our energy needs. A wave generator is in two sections hinged together so they can move relative to each other, thus turning a generator in one of the sections. The incoming waves make the two sections rock relative to each other. Scientists reckon that about 1 kilometre of coastline could produce about 50 megawatts of electrical power.

Insight

Tidal power and hydroelectric power stations are much more reliable than wind turbines and wave generators. Use a Web search engine to locate the biggest hydroelectric power station and the biggest tidal power station in the world.

QUESTIONS

1 For each type of heat engine listed below, what is (i) the source of
   the high temperature, (ii) the low-temperature sink?
   (a) An internal combustion engine.
   (b) A jet engine.
   (c) A steam engine.

2 Calculate the efficiency of an internal combustion engine in which
   air is heated to 650 K when the surrounding temperature is
   (a) 300 K, (b) 320 K.

3 Britain's electricity demand in winter is about 60 000 megawatts.
   About one-half of Britain's electricity is provided by oil- and
   gas-fired power stations, about a third is provided by coal-fired
   power stations and about a sixth is produced by nuclear power
   stations.
   (a) How many megawatts of electricity is provided by these
        nuclear power stations?
   (b) How many megawatts of electricity is provided by fossil fuel
        power stations?
   (c) State one advantage and one disadvantage of (i) a nuclear
        power station, (ii) a coal-fired power station, (iii) a gas-fired
        power station.

4 Britain's oil and gas reserves will probably run out by 2050. Suggest
   possible strategies that could be developed to make up the 'energy
   gap' after these reserves have been used up.

5 (a) A supplier of solar heating panels estimates that a panel of
        area 1 square metre absorbs 500 W of solar energy on a sunny
        day. Calculate how many of these panels would be needed to
        supply 3000 W of solar energy on such a day.
   (b) A wind turbine is capable of supplying 2 MW of electrical
        power. How many such wind turbines would be needed to
        supply the same amount of electrical power as a 5000 MW
        nuclear power station?

(c) *Wave generators could provide 50 MW of electrical power for every kilometre length of coastline. What length of coastline would need to be used to generate the same amount of electrical power as a 5000 MW nuclear power station?*

10 THINGS TO REMEMBER

1 *A heat engine needs a low-temperature 'sink' as well as a high-temperature 'source' to do work.*

2 *The* **maximum efficiency** *of a heat engine operating between high temperature T_1 and low temperature T_2*

$$= \frac{T_1 - T_2}{T_1}$$

3 *The* **internal energy** *of an object is the energy it possesses, not counting its kinetic or potential or elastic energy.*

4 *The* **First Law of Thermodynamics** *states that when an object or a system of objects does work or gains heat, the change of internal energy of the object(s) is equal to the difference between the heat gained and the work done by the object(s).*

5 *The* **Second Law of Thermodynamics** *states that energy tends to spread out and become less useful whenever it is transferred between objects in an isolated system.*

6 **Fossil fuels** *include coal, oil and gas.* **Non-fossil fuels** *include wood and uranium.*

7 **Renewable energy resources** *include hydroelectric and tidal power stations, wind turbines, wave generators and solar panels.*

8 *Nuclear power stations produce* **radioactive waste** *which must be stored securely for many years.*

9 *Fossil fuel power stations produce* **greenhouse gases** *which are causing global warming.*

10 *Renewable energy resources such as wind farms and tidal power stations need to cover large areas.*

7

Electricity

In this chapter you will learn:
- *how electricity was discovered*
- *how current electricity was first investigated by Ampère*
- *about the nature and basic principles of electricity*
- *how electric circuits transfer energy*
- *how electricity is used at home.*

Electricity provides most of the power we use at home and in the workplace. Before the electricity distribution system was established, people used gas or solid fuels for heating and cooking. Lighting was provided by gas or oil lamps or candles. Today electrical appliances are indispensable, as anyone who has suffered an electrical power failure knows. In many underdeveloped countries, electricity does not reach remote villages and consequently most people live in poverty. The principles that underpin electricity generation and distribution were discovered in the nineteenth century, and electricity distribution systems were set up in the late nineteenth century and the early decades of the twentieth century. In the future, renewable energy sources such as solar panels and wind turbines will probably provide electricity without causing pollution problems in urban regions as well as providing electricity in remote regions. In this chapter, we will look in depth at the principles of electricity and how electricity is generated and distributed.

The nature of electricity

Any material that allows electricity to pass through it is known as an **electrical conductor**. All metals and certain non-metals, such as graphite, conduct electricity. Materials that do not allow electricity to pass through them are known as **electrical insulators**. Examples of electrical insulators include polythene, nylon, Bakelite, air and oil. An electric torch consists of a battery connected to a torch bulb, usually with one terminal of the battery connected directly to the torch bulb and the other terminal connected to the torch bulb via a switch. When the switch is closed, the torch bulb lights because the two metal parts of the switch make contact. When the switch is opened, the two metal parts move apart and the bulb goes off. A complete circuit of electrical conductors is needed for electricity to pass through the torch bulb. If there is a gap in the circuit, no electricity can pass round the circuit.

STATIC ELECTRICITY

Static electricity is produced in a thunderstorm when clouds become charged with electricity. When charged clouds can hold no more electricity, they discharge to Earth in massive lightning strokes. A less dramatic demonstration of static electricity can be achieved by rubbing an inflated balloon with a dry cloth. You may be able to feel tiny sparks of static electricity discharging to your hand if you then touch the charged balloon. The charged balloon will attract bits of paper and will stick to a ceiling. Certain insulating materials, such as glass, polythene, perspex, nylon and rubber, also become charged when rubbed with a dry cloth.

Insight

Tiny sparks can be catastrophic where dry powders or flammable liquids and vapours are present. The disastrous explosion in 2005 at the Buncefield oil depot near London is thought to have been set off by a spark. The explosion was reportedly heard as far away as Holland. Key 'Buncefield' into your Web search engine to read eyewitness reports.

A charged object exerts a force on any other charged object. For example, two charged polythene rods repel each other when held close together. This may be demonstrated by charging the end of one of the rods and suspending it horizontally on a thread. The other rod is then charged at one end and repels the end of the suspended rod when placed close to it. The same effect is observed if two charged Perspex rulers are held close together. However, if a charged Perspex ruler is held near a charged polythene rod, the two objects attract each other. These tests show that:

1 *there are two types of electric charge*
2 *two objects that carry like charge (i.e. the same type of charge) repel each other*
3 *two objects that carry unlike charge (i.e. different types of charge) attract each other.*

The two types of charge are referred to as **positive** and **negative charge** because they cancel each other out if one type is brought into contact with the other type. A charged polythene rod carries a negative charge. Therefore, any charged object that is repelled by a charged polythene rod carries a negative charge. A charged Perspex ruler carries a positive charge and therefore it attracts a charged polythene rod.

The action of rubbing such a material charges it with electricity because tiny negatively charged particles transfer between the material and the cloth used to rub it. These charged particles are called **electrons** and they are in every atom of every substance. Every atom contains a positively charged nucleus surrounded by electrons. An uncharged atom has the same amount of negative charge as positive charge.

▶ *If electrons are added to an uncharged atom, the atom becomes negatively charged because it now contains more negative charge than positive charge.*
▶ *If electrons are removed from an uncharged atom, the atom becomes positively charged because it now contains more positive charge than negative charge.*

Figure 7.1 Charging by friction.

CURRENT AND CHARGE

An electric current is a flow of charge, usually carried by electrons. The electrons in an insulator are trapped inside the atoms, but in a conductor some of the electrons are not trapped and can move about freely inside the conductor. A torch lamp circuit consists of a cell or a battery (two or more cells) connected to the torch bulb via two wires and a switch. When the switch is closed, electrons pass round the circuit because all the parts of the circuit can conduct electricity. The battery forces the free electrons in the conductors one way round the circuit through the wires, the torch bulb, the switch and the battery.

Figure 7.2 Electrons at work.

When a circuit is switched on, electrons leave the negative terminal of the battery, move round the circuit and re-enter the battery at its positive terminal. The one-way flow of charge round an electric circuit was first deduced by the French scientist, Andre Ampère. Before Ampère's discovery, it was thought that positive charge flowed from the positive terminal of the battery and was cancelled out by negative charge flowing from the negative terminal. Ampère knew that an electric current along a wire caused a magnetic compass near the wire to deflect and he knew that this was because a magnetic field was created around the wire when an electric current passed along it. He also noticed that the compass deflected in the opposite direction if the battery was reversed in the circuit. He realized that this observation could only be explained if only one type of charge flows round a circuit. Reversing the battery caused the flow of charge in the circuit to reverse, which reversed the direction of the magnetic field round the wire. However, Ampère was not able to say if the charge was positive charge from the positive terminal or negative charge from the negative terminal of the battery. To prevent confusion, scientists agreed that the direction of current in a circuit should be the direction of flow of positive charge. We know now that an electric current in a wire is a flow of electrons round the circuit from the negative terminal to the positive terminal of the battery. Nevertheless, the rule that the direction of current is the direction of flow of positive charge is still used today.

The **unit of electric current** is the **ampere** (A), defined in terms of its magnetic effect, and can be measured using an ammeter. A current of 1 A passing along a wire is due to 6.25 million million million electrons passing along the wire every second. Each electron carries the same amount of negative charge. The bigger the current in a wire, the greater the number of electrons passing along the wire each second. Therefore, the charge passing along a wire in a certain length of time is proportional to the current and the duration of the time interval.

Note: Small currents are measured in *milliamperes* (mA) or *microamperes* (µA), where 1 mA = 0.001 A and 1 µA = 0.001 mA.

The **unit of charge,** the **coulomb** (C), is defined as the charge that passes along a wire in 1 second when the current is 1 ampere. Therefore, for a wire which carries a constant current,

> The charge passing in a certain interval of time = current × time interval:
>
> Charge (in coulombs) = current (in amperes) × time (in seconds).

For example, the charge passing along a wire

- ▶ *for a current of 2 amperes in 5 seconds is 10 coulombs*
- ▶ *for a current of 2 amperes in 20 seconds is 40 coulombs*
- ▶ *for a current of 20 amperes in 5 seconds is 100 coulombs.*

Each electron carries a negative charge so small that 6.25 million million million are needed to make up 1 coulomb of charge. 6250 million million electrons pass through a 1 mA torch bulb every second when it lights normally.

BATTERIES AND CELLS

In any circuit, electrons transfer energy from the battery to the components in the circuit. Each electron in an electric circuit gains energy as it passes through the battery, leaves the battery via the negative terminal, loses energy as it passes round the circuit and re-enters the battery via its positive terminal. Each electron leaves the negative terminal of the battery with electrical potential energy, which is used up and converted to other forms of energy as the electron passes round the circuit. For example, when an electron passes through a torch bulb, energy is transferred from the electron to the torch bulb. The effect of all the electrons passing through the torch bulb filament is to make it so hot that it emits light. The connecting wires are good conductors so no energy is given up in these wires, provided the current is not excessive.

A **battery** consists of two or more identical **cells,** each cell consisting of two electrodes in a conducting paste or liquid, the electrolyte, which reacts with the electrodes. The two electrodes are made from different materials such as graphite and lead, chosen because they react differently with the electrolyte. One electrode loses electrons to the electrolyte and the other one gains electrons from the electrolyte. Thus one electrode becomes positively charged and the other one becomes negatively charged. In use, a cell gradually deteriorates because the chemical reactions between the electrodes and the electrolyte convert the electrolyte and the electrodes into other substances which do not react. In a rechargeable cell, this chemical process can be reversed by connecting the cell to a battery charger which forces electrons into the cell at the negative electrode and out at the positive electrode. Disposable batteries and cells cannot be recharged and are discarded when exhausted.

The **voltage** of a battery or a cell is the power in watts that the cell can deliver to the rest of the circuit for every ampere of current passing through it. Thus a 12 volt battery connected in a circuit delivers 12 watts of power for every ampere of current passing through it. If it is connected to a suitable 3A light bulb, the battery supplies 36 watts of power to the light bulb because the current is 3 amperes.

Voltage (in volts) = power delivered (in watts) per ampere of current:

$$\text{Voltage or potential difference (in volts)} = \frac{\text{power delivered in watts}}{\text{current in amperes}}$$

The capacity of a battery or cell is usually expressed in ampere hours. This is the number of hours for which it would be able to supply 1 ampere of current before it is exhausted. Thus a car battery with a capacity of 60 ampere hours would be able to supply a current of 1 ampere for 60 hours. Such a battery could supply a current of 2 amperes for 30 hours or a current of 5 amperes for 12 hours.

Cold start

One reason why cars sometimes fail to start in winter is because a bigger current than normal is needed to start a very cold engine. The oil in the engine becomes more sluggish as its temperature falls and so the moving parts are more difficult to move, the colder the engine is. If the battery is not fully charged, it fails to start the engine because it cannot supply enough current. Switching on the car lights and the heater before starting the car would make starting even more difficult.

MORE ABOUT VOLTAGE

In an electric circuit that contains a battery, current passes round the circuit and through the battery, transferring energy at a steady rate from the battery to the rest of the circuit. The energy supplied every second by the battery is delivered to the rest of the circuit. A battery in an electric circuit acts like a pump in a central heating system; the pressure of the pump forces water through one radiator after another round the system. There is a pressure drop across each radiator because the pressure at the inlet to each radiator is higher than at the outlet. The sum of the pressure drops round the system is therefore equal to the pump pressure. The pressure drop across each radiator is a measure of the work done by the pump to force the water through the radiator. Thus the sum of the pressure drops round the system is a measure of the total work done by the pump, which is measured by the pump pressure. In the same way, the battery voltage is a measure of the power supplied by the battery, which is equal to the power delivered to all parts of the circuit. Hence the battery voltage is equal to the sum of the voltages round the circuit.

The voltage between any two points in an electric circuit is the power delivered per ampere of current to that part of the circuit between the two points.

For example, if the voltage between two points in a circuit was 20 volts, then every ampere of current passing from one point to the other would deliver 20 watts of power. Thus a current of 5 amperes passing from one point to the other would cause 100 watts (= 20 volts × 5 amperes) to be delivered to that part of the circuit.

Insight

A battery in a circuit forces electrons round the circuit, delivering energy from the battery to the rest of the circuit. The voltage across a component in the circuit is a measure of the energy delivered by each electron passing through it.

Voltage or potential difference (in volts) = power delivered in watts per ampere of current.

QUESTIONS

1 Select the correct words from the following list to complete the passage below.

negative   positive   gains   loses

A polythene rod becomes negatively charged when it is rubbed with a dry cloth. This is because electrons carry a _____ charge and the rod _____ electrons and the cloth _____ electrons when they are rubbed together.

2 (a) State the unit of electric charge.
   (b) How much electric charge passes through a torch bulb when the current is 0.030 A for (i) 10 seconds, (ii) 5 minutes?

3 A 12 volt car battery is connected to a 12 V 3 A light bulb. The car battery has a capacity of 36 ampere hours. Calculate:
   (a) how long the battery can continue to supply this amount of current before it becomes flat,
   (b) how much power the battery supplies to the light bulb.

Electric circuits

CIRCUIT RULES

A **series circuit** is one in which the same current passes through all
the components in the circuit, as shown in Figure 7.3 (a). In other
words, all the electrons moving round such a circuit pass through
every component in the circuit. The current may be measured using an
ammeter, which is a meter designed to measure current. The ammeter
must be connected in series with the other components in the circuit so
that the same current passes through all the components in the circuit.

A circuit with components **in parallel** with each other is one in
which part of the current from the battery passes through each of the
parallel components. See Figure 7.3 (b). Components in parallel in
a circuit have the same voltage between the two points where they
are connected to the circuit. Each electron moving round the circuit
passes through one component or the other component when it moves
from one point to the other. The voltage between two points in a
circuit may be measured using a **voltmeter,** which is a meter designed

Figure 7.3 (a) Components in series. (b) Components in parallel.

to measure voltage. To measure the voltage across a component, the voltmeter must be connected in parallel with the component.

▶ *For components in series, the* **current** *is the same.*
▶ *For components in parallel, the* **voltage** *is the same.*

RESISTANCE

A potential difference or voltage exists between two points in an electric circuit if an electron has to do work to move from one point to the other. Conductors in a circuit resist the flow of electrons because the electrons repeatedly collide with the atoms of the conductor and lose energy in these collisions. The atoms of the conductor vibrate and the vibrations increase when the atoms gain energy from the electrons which collide with them. As a result, the conductor gains energy which it loses to the surroundings through heat transfer. In effect, the conductor resists the flow of electrons through it and the electrons must use some of their energy to pass through the conductor. Thus if a voltage exists between two points in any circuit, the electrons must use some or all of their energy to overcome the resistance of that part of the circuit to their passage from one point to the other.

The **resistance** of a circuit component is defined as

$$\frac{\text{the voltage across the component}}{\text{the current through the component}}$$

The unit of resistance is the **ohm** (symbol Ω), which is the amount of resistance between two points in a circuit when the voltage between the two points is 1 volt and the current is 1 ampere.

$$\text{Resistance (in ohms)} = \frac{\text{voltage (in volts)}}{\text{current (in amperes)}}$$

Note: The resistance formula above can be rearranged as:

$$\text{Voltage} = \text{current} \times \text{resistance} \quad \text{OR} \quad \text{Current} = \frac{\text{voltage}}{\text{resistance}}$$

Worked example

Calculate the resistance of a wire when the current through it is 2.0 A and the voltage across its ends is 3.0 volts.

SOLUTION

$$\text{Resistance} = \frac{\text{voltage}}{\text{current}} = \frac{3.0 \text{ volts}}{2.0 \text{ amperes}} = 1.5\,\Omega$$

CIRCUIT DIAGRAMS

A light bulb	H voltmeter
B cell	I variable resistor
C resistor	J light-dependent resistor
D fuse	K thermistor
E switch	
F diode	
G ammeter	

Figure 7.4 Electrical symbols.

Standard symbols are used to represent components in circuit diagrams. The function of each component represented in Figure 7.4 is described below, if it has not been previously described.

▶ A **resistor** *is a component designed to have a known resistance. For example, to make a 5.0 Ω wire-wound resistor from wire of resistance 2.5 ohms per metre, a length of exactly 2.0 metres would need to be used. The current through a resistor is constant for a fixed voltage across the resistor. The bigger the resistance of the resistor, the smaller the current is.*

▶ A **variable resistor** *is a component designed to change the current without necessarily changing the voltage. A variable resistor can be made by changing the length of a resistance wire in a circuit. If the length is increased, the resistance of the wire is increased and so the current becomes smaller.*

▶ A **diode** *allows current through in one direction only. A diode conducts when it is connected into the circuit in its forward direction, as its resistance in the forward direction is very low. If the diode is then reversed, it does not conduct because its resistance in the reverse direction is very high.*

▶ A **light-emitting diode** *(LED) emits light when current passes through it. LEDs are used as indicators in electronic circuits.*

Insight

A **superconductor** is a material with zero resistance. Such a material needs to be below a certain temperature, its **critical temperature,** to be superconducting. The highest critical temperature known at present at atmospheric pressure is −135°C for a metal oxide containing 5 different metals including copper.

THE ELECTRICAL POWER EQUATION

Because the voltage between any two points in a circuit is the power delivered per ampere, it follows that the power delivered for a certain current can be calculated by multiplying the voltage by the current. In other words,

> Power delivered (in watts) = voltage (in volts) × current (in amperes).

For example, an electric kettle that operates at a voltage of 230 volts and a current of 10 amperes would take 2300 watts of power (= 230 volts × 10 amperes) from the electricity supply when it is switched on. A car's electrical heater designed to operate at 12 volts and 5 amperes would take 60 watts of power (= 12 volts × 5 amperes) from the car battery, if the engine was not switched on.

QUESTIONS

4 *A current of 2.5 A is passed through a 6.0 V resistor. Calculate*
   (a) the voltage across the resistor, (b) the power supplied to the resistor.

5 *(a) Draw a circuit diagram to show a diode connected in series with a 1.5 V cell and a lit torch bulb.*
   (b) If the 1.5 V cell was connected in the circuit in the reverse direction, the torch bulb would not light. Why?

6 *A set of Christmas tree lights contains 20 light bulbs in series, each rated at 12 volts and 6 watts.*
   (a) Calculate (i) the current through the light bulbs, (ii) the resistance of a single light bulb.
   (b) Explain why none of the lights will light up if any one of the light bulbs fails.

Electricity at work

A magnetic compass near a wire is deflected when the current in the wire is switched on. This is because the current creates a magnetic field round the wire. With no current in the wire, the compass needle points north because of the Earth's magnetic field. If the compass was moved in the direction its needle points, it would follow a straight line northwards. Such a line is called a magnetic field line, or sometimes a magnetic line of force. Its direction is defined as the direction in which a compass needle points. With the current on, the compass would need to move round the wire because the magnetic field lines are circles centred on the wire, as shown in Figure 7.5. If the current is reversed in the wire, the direction of the magnetic field lines round the wire reverses.

clockwise concentric magnetic field lines   anticlockwise concentric magnetic field lines

Figure 7.5 The magnetic field around a current-carrying wire.

ELECTROMAGNETS

An electromagnet consists of a coil of insulated wire wound round an iron bar. When a current is passed through the coil, the iron bar is magnetized by the magnetic field due to the current. As a result, the iron bar is able to attract iron and steel objects held near its ends. When the current is switched off, the iron bar loses its magnetism. Electromagnets have many uses, from powerful electromagnets used to pick up and move cars in scrapyards to very sensitive electromagnets used to write data onto computer discs. Two further uses of electromagnets are described below.

▶ **The relay:** *when current is passed through the coil of a relay, the electromagnet attracts an iron armature. The movement of this armature opens or closes a switch which is part of a different circuit. When the current is switched off, the armature springs back to its normal position and the switch reverts to its original state.*

Figure 7.6 (a) The relay. (b) The electric bell.

▶ **The electric bell:** *the electromagnet coil is part of a 'make-and-break' switch. When current passes through the coil, the electromagnet attracts the iron armature, which makes the hammer hit the bell. The movement of the armature opens the 'make-and-break' switch which switches the electromagnet off, allowing the armature to spring back and close the switch. Current then passes through the electromagnet and the sequence is repeated.*

ELECTRIC MOTORS

A wire near a magnet experiences a force when a current is passed along the wire. This effect is known as the **motor effect**. The reason for the effect is that the electrons passing along the wire are pushed sideways by the magnet, exerting a sideways force on the wire as a result. The effect depends on the angle between the wire and the lines of the magnetic field. For maximum effect, the wire needs to be at right angles to the lines of the magnetic field.

An electric motor consists of a coil of insulated wire which spins between opposite poles of a U-shaped magnet or electromagnet. When current is passed round the coil, the coil turns because the wires along one edge of the coil are forced up and the wires along

(a)

force

current

Current passes along each side of the coil in opposite directions. Each side is therefore acted on by a force due to the magnetic field. The force on one side is in the opposite direction to the force on the other side.

(b)

coil

force

force

split-ring commutator

When the coil is parallel to the field, the forces on the sides rotate the coil. As the coil turns through the position at 90° to the field, the split-ring commutator reverses its connections to the battery, reversing the current direction round the coil.

Rotation by more than 90° has swapped the positions of the two sides and reversed the current direction in the coil. Therefore, the forces acting on each side continue to turn the coil in the same direction as before, so the coil rotates continuously in one direction.

Figure 7.7 (a) The motor effect. (b) The electric motor.

the opposite edge are forced down. After half a turn, the current round the coil must be reversed, otherwise the coil is turned back by the magnetic field. This reversal is achieved automatically by the action of the split-ring commutator which rotates with the coil. Conducting brushes made of graphite connected to the battery press against the commutator. Graphite is used as it is a conductor and provides contact with little friction.

The split-ring commutator has two functions:

1 *It provides continuous contact between the battery and the coil as the coil turns.*
2 *It reverses the current direction round the coil every half turn, so the coil is forced to spin in the same direction every half turn. Without this reversal of the current direction every half turn, the coil would reverse its direction repeatedly.*

If the battery is reversed in the circuit, the motor spins in the opposite direction. The greater the current, the faster the rate of rotation of the coil. In a mains electric motor:

▶ *an electromagnet is used instead of a permanent magnet*
▶ *a number of coils at equal angles are wound on an iron core, thus providing smoother rotation. Each coil is connected to opposite sections of a multi-segment commutator.*

GENERATORS

Electricity can be generated in a coil by moving the coil in a magnetic field. Provided the coil wires cut the lines of the magnetic field, a voltage is induced in the coil. If the coil is part of a complete circuit, the induced voltage makes a current pass round the circuit. A cycle **dynamo** consists of a magnet which is forced to rotate near the end of a coil, causing a voltage to be induced in the coil. The faster the magnet rotates, the greater the voltage.

A voltage is induced in any wire in a magnetic field provided that the wire and magnet are moving relative to each other and the lines of the magnetic field cut across the wire. This effect was discovered by Michael Faraday in 1831. The cause of the effect is that the electrons in the wire move through the field as the wire moves. As a result, the electrons in the wire are pushed by the magnetic field to one end of the wire. This push is sometimes referred to as an electromotive force and it exists only as long as the wire and magnet move relative to each other, so that the wire cuts across the magnetic field.

Insight
'Induced' means brought about'. An induced voltage in a wire is brought about when the wire cuts the lines of the magnetic field.

The **alternating current generator** consists of a rectangular coil of insulated wire that is forced to rotate between the poles of a U-shaped magnet. The wires along opposite edges of the coil cut across the magnetic field lines as the coil rotates, causing a voltage to be induced in the coil. The coil wires are connected to two slip rings on the axle of the coil. A graphite 'brush' pressed against each slip ring provides a continuous contact to an external circuit.

The voltage reverses in polarity as each edge of the coil crosses the middle from one magnetic pole to the other pole every half turn. The voltage peaks when the coil edges are nearest the poles. Thus the voltage varies as shown in Figure 7.8, acting in each direction every half turn. This type of voltage is referred to as an **alternating voltage**. The voltage is said to 'alternate' because it repeatedly reverses its direction. If the generator is connected to a suitable resistor, an alternating current is forced round the circuit. For each complete turn of the coil, the current reverses and reverses back again in a full cycle.

Figure 7.8 (a) Electromagnetic induction. (b) The a.c. generator.

- ► The **frequency** *of an alternating current or voltage is the number of cycles that take place each second. The unit of frequency is the hertz (Hz), where 1 hertz is equal to 1 cycle per second.*
- ► The **peak voltage** *of an alternating current or voltage is the maximum voltage or current in either direction.*

TRANSFORMERS

A transformer is designed to step an alternating voltage up or down. It consists of two coils of insulated wire, referred to as the primary and secondary coils, wound round the same iron core. When an alternating voltage is applied to the primary coil, an alternating voltage is induced in the secondary coil. The reason is that an alternating current passes through the primary coil and creates an alternating magnetic field in the core. This alternating magnetic field induces an alternating voltage in the secondary coil, just as a rotating magnet in a dynamo induces a voltage in the coil of the dynamo.

Figure 7.9 The transformer.

The ratio of the secondary voltage to the primary voltage is equal to that of the number of windings or 'turns' on the secondary coil to the number on the primary coil.

$$\frac{\text{Secondary voltage}}{\text{Primary voltage}} = \frac{\text{number of secondary turns}}{\text{number of primary turns}}$$

▶ *A step-up transformer has more windings on the secondary coil than on the primary coil. For example, if there are 20 times as many secondary turns as primary turns, the secondary voltage is 20 times the primary voltage.*

▶ *A step-down transformer has fewer windings on the secondary coil than on the primary coil. For example, if there are 20 times fewer secondary turns than primary turns, the secondary voltage is a twentieth of the primary voltage.*

THE ELECTRICITY GRID

The electricity grid consists of a network of cables and transformers connecting power station generators to mains electricity users. Power is wasted in a cable if the current is too large, because of the resistance of the cable. Because electrical power = current × voltage, the same amount of electric power can be delivered using low current and high voltage as using high current and low voltage. Power station generators are connected to the grid system by means of step-up transformers. The high voltage on the cables means that the current through the cables is too low to heat the cables and so little power is wasted. Step-down transformers at local sub-stations are used to reduce the voltage to a suitable level to match the needs of the user. Electricity supplied to homes and offices is at 230 volts.

QUESTIONS

7 (a) *State two uses of an electromagnet.*
   (b) *With the aid of a diagram, explain the operation of an electric bell.*

8 (a) *Draw a labelled diagram of a simple electric motor and use your diagram to explain how the electric motor works.*

(b) State and explain what happens if the battery connected to a simple electric motor is (i) reversed, (ii) replaced by an alternating voltage supply.

(c) A mains electric motor has an electromagnet instead of a permanent magnet. The electromagnet and the armature coil are both connected to the mains, which supplies alternating voltage. Explain why this type of motor spins in one direction only, regardless of the direction of the current through the coil.

9 (a) With the aid of a diagram, describe the operation of an alternating current generator.

(b) A power station generator supplies electricity at 11 000 volts via a step-up transformer to the grid system at 132 000 volts.
  (i) Calculate the turns ratio of the transformer.
  (ii) The generator supplies 400 kilowatts of power to the grid. Calculate the current (1) through the generator circuit, (2) through the secondary coil of the transformer.
  (iii) Explain why less power is wasted in the grid system by transmitting electricity at high voltage rather than low voltage.

Electricity in the home

Mains electricity is supplied from a local sub-station via a main cable, usually underground. The cable consists of two low-resistance insulated conductors. When a mains appliance is switched on, the appliance becomes part of a complete circuit consisting of the two conductors and the secondary coil of the sub-station transformer. One of the two conductors in the main cable, the **neutral wire**, is connected to the ground (i.e. 'earthed') at the sub-station so its voltage is zero. The other conductor, the **live wire**, alternates in voltage between a maximum positive and a maximum negative voltage. In most countries except North America and parts of South America, the mains voltage alternates between +340 V and −340 V.

The voltage of the mains is usually measured and stated in terms of the value of the steady (i.e. direct) voltage which would deliver the same power on average to any mains appliance. For a mains voltage that alternates between +340 V and –340 V, the 'equivalent' steady voltage works out at about 230 volts. Thus a 230 volt 1000 watt mains electric heater would provide the same heating effect if connected to a 230 volt direct voltage supply as it would if connected to 230 volt mains.

Safety first

The live wire and any terminals connected to it are dangerous. Anyone who touches a live wire would suffer a fatal electric shock because the human body conducts electricity. No more than about 0.02 amperes through the body would cause a severe electric shock. The resistance of the body is about 1000 ohms. Voltages in excess of about 20 volts (= 0.02 amperes × 1000 ohms) are therefore dangerous. The live wire reaches about 340 volts every half cycle.

DOMESTIC CIRCUITS

The mains cable is connected to the circuits in a house via the distribution board. A main switch can be used to cut the electricity supply off at the distribution board. A fuse in each circuit at the distribution board cuts off the current to any circuit in the event of a fault. Each circuit in a house consists of insulated live and neutral wires connected to the live and neutral terminals of the distribution board.

▶ *Lighting circuits in the house usually consist of low-resistance cables. Individual sockets connected to the same cable are connected in parallel with one another.*

Each circuit from the fuse board is protected with its own fuse. If the fuse 'blows', the live wire is therefore cut off from appliances supplied by the circuit.

The mains cable from the substation to a building is connected via the electricity meter to the circuits in the building at the distribution fuse board. The live wire from the substation is connected via a main fuse to the electricity meter.

The two wires used to supply an electric current to an appliance are referred to as the **live** and the **neutral** wires. The neutral wire is earthed at the nearest mains substation.

Mains wires need to have as low a resistance as possible, otherwise heat is produced in them by the current. This is why mains wires are made from copper. All mains wires and fittings are insulated.

The fuse in a lighting circuit is in the fuse box. Each light bulb is turned on or off by its own switch. When the switch is in the off position, the appliance is not connected to the live wire of the mains supply.

A ring main is used to supply electricity to appliances via wall sockets. A ring main circuit consists of a live wire, a neutral wire and the earth wire which is earthed at the fuse board. The wires of a ring main are thicker than the wires of a lighting circuit because appliances connected to a ring main require more current than light bulbs do.

Each appliance is connected to the ring main by means of a three-pin plug which carries a fuse. An appliance a metal chassis is earthed via the three-pin plug and the earth wire. This prevents the metal chassis from becoming live if a fault develops in the appliance. Appliances connected to the ring main can be switched on or off independently since they are in parallel with each other.

Figure 7.10 The ring main.

- The ring main circuit in a house supplies electricity to the wall sockets via insulated live and neutral wires. The ring main cable also includes a third insulated wire, the earth wire, which connects the metal chassis of any electrical appliance connected to the ring main to earth at the house. This is to prevent the metal chassis from becoming live if a live wire becomes loose in the appliance.
- The sockets of the ring main are connected in parallel with each other. Each plug used to connect an appliance to the ring main is fitted with its own fuse. The switch in each socket is always connected on the 'live side' of the socket. More current can be supplied via the ring main than via lighting circuits because the wires are thicker and therefore conduct better. Also, the ring main provides two routes for current from the distribution board for any appliance.

FUSES

A fuse is a thin wire designed to melt if the current through it exceeds a certain value referred to as the **fuse rating**. If a fuse melts, no current can pass through it because there is then a gap in the circuit. Thus a fuse is an essential safety feature of an electric circuit, intended to cut the current off if the current becomes too large. For example, an electric kettle fitted with a 13 A fuse operates normally when switched on provided that the current does not exceed 13 A. If a fault in the kettle or the connecting cable causes the current to exceed 13 A, the fuse melts and cuts the current off.

Safety first again!

Always use a fuse of the recommended rating for a given appliance. For example, if a 5 A fuse is recommended for a hairdryer, a 13 A fuse would be unsafe as it would not melt if current exceeded the maximum safe value of 5 A, and a 3 A fuse would melt every time the hairdryer was switched on to full power.

ELECTRICITY COSTS

Electricity meters measure electricity in units of kilowatt hours, where 1 kilowatt hour is defined as the energy used by a 1 kilowatt (= 1000 watts) appliance in 1 hour. A 3 kilowatt electric heater would therefore use 3 units (i.e. 3 kilowatt hours) of electricity in one hour. A 100 watt light bulb would use 2.4 units of electricity if switched on for 24 hours.

If an electricity bill shows that 2500 units of electricity were used in a certain period, this means that the energy used was the same as a 1 kilowatt appliance would use in 2500 hours. An electricity bill also states the unit cost of electricity, which is the cost per kilowatt hour. Thus the cost of 2500 units of electricity at 10p per unit would be £250 (= 2500 × 10p).

Worked example
The electricity usage in a household was monitored for one week. During that time, the electricity usage was as follows.

A 3 kilowatt electric kettle was used 20 times for 6 minutes each time (i.e. 2 hours total).

An 800 watt microwave cooker was used 6 times for 10 minutes each time.

A 500 watt television was used for a total of 10 hours.

Four 100 watt electric lights were used for a total of 40 hours.

A 2 kilowatt electric heater was used for a total of 20 hours.

(a) Calculate the number of units of electricity used by each appliance above.

(b) Calculate the total cost of the electricity supplied above for a unit cost of 10p per unit.

SOLUTION

(a) Kettle = 3 kW × 2 hours = 6 units;
Microwave cooker = 0.800 kW × 1 hour = 0.8 units;
Television = 0.500 kW × 10 hours = 5 units;
Lighting = 4 × 0.100 kW × 40 hours = 16 units;
Heater = 2 kW × 20 hours = 40 units.

(b) Total number of units used = 6 + 0.8 + 5 + 16 + 40 units =
67.8 units

Cost = 67.8 × 10p = £6.78p.

QUESTIONS

10 A 1000 watt 230 volt electric heater is to be fitted with a
fuse. Which one of the following fuses should be chosen:
3 A, 5 A, 13 A?

11 (a) State the purpose of the earth wire in a ring main circuit.
(b) Why are the wires used for ring main circuits thicker than the
wires used in lighting circuits?
(c) A short circuit occurs if a fault develops and a live wire or
terminal makes contact with an earthed or neutral wire.
Explain why (i) there is a risk of fire in a short circuit, and
(ii) why a correctly fitted fuse prevents such a fire risk.

12 An electricity meter reading was 28 501 units on a certain day and
28 642 units one week later.
(a) Calculate the cost of the electricity used in this period, given
each unit cost 11p.
(b) In the period above, a 6 kilowatt electric water heater was
used for a total of 16 hours. Calculate how many units were
due to the heater and the percentage of the total cost due to
the heater.

10 THINGS TO REMEMBER

1 *There are two types of electric charge,* **positive** *and* **negative**.

2 *Like charges repel; unlike charges attract.*

3 *An* **electric current** *is a flow of charge and is measured in amperes.*

4 *The* **voltage** *(in volts) across a component is the power (in watts) delivered to it per ampere of current in it.*

5 *Components* **in series** *have the same current; components* **in parallel** *have the same voltage.*

6 **Resistance** *(in ohms) = voltage (in volts)/current (in amperes).*

7 **Electrical power** *(in watts) = voltage (in volts) × current (in amperes).*

8 *The* **motor effect** *is the force on a current-carrying wire at a non-zero angle to the lines of a magnetic field.*

9 *An* **alternating current generator** *consists of a coil that spins in a magnetic field.*

10 *A* **transformer** *is used to step an alternating voltage up or down.*

8

The nature of light

In this chapter you will learn:

- *about the properties of light, including reflection, refraction and interference*
- *why Newton's theory of light was eventually rejected in favour of wave theory*
- *about electromagnetic waves and their properties*
- *what photons are and how Einstein established the photon theory of light.*

In this chapter, we will start by studying reflection and refraction of light and the formation of images. We'll then move on to consider two different theories put forward over 300 years ago about the nature of light, both of which can explain reflection and refraction. One theory said that light was made up of tiny particles. The other theory said that light consists of waves. Which theory proved to be correct? We will look at the evidence for and against each theory and how, eventually, a century later, more evidence was obtained which led to the acceptance of one theory and rejection of the other.

The story of the nature of light had further twists in store with the discovery of photoelectricity at the end of the nineteenth century. Investigations into this effect produced observations that could not be explained until a new theory of light, namely the photon theory, was proposed by Albert Einstein, then a young physicist employed by the Swiss Patent Office. Einstein went on to develop his theories of relativity, which include the astonishing outcome that nothing can travel faster than light. The **speed of light** in space at 300 000 kilometres per second is the cosmic speed limit. Even more astounding is Einstein's discovery that energy and mass are interchangeable on a scale according to his famous equation $E = mc^2$, where c is the speed of light. Our understanding of light has developed over several centuries and continues to develop as physicists strive to understand why light provides the link between energy and mass. So let's look at this story in more detail.

Properties of light

A laser beam or a beam of sunlight travels in a straight line. On a cloudy day, sunlight that breaks through a gap in the clouds can be seen as a straight beam extending to the ground. We talk about the Sun's rays, but what do we mean by the word 'ray'? It seems a very convenient word for the path that light takes when it radiates from a source (e.g. light from the Sun). So for convenience we will think of light in terms of rays. Also, let us assume for now that light travels in straight lines.

REFLECTION OF LIGHT

Stand in front of a flat mirror and you will see an image of yourself. You will also be able to see beyond your own image the image of any other object behind you. Each image is the same distance behind the mirror as the object is in front. So if you stand 50 centimetres in front of a flat mirror, your image is 50 centimetres behind the mirror and is therefore 100 centimetres away from you. Step back from the mirror and your image steps back by the same distance.

Mirror images
Make a semi-transparent 'mirror' from cellophane film over an open box. Put the box on its side so the film is vertical and place

a small object in front of it. You should be able to see an image of the object in the box. Move the object about until its image appears to be at the back of the box. You should find the distance from the object to the 'mirror' is then the same as the distance from the 'mirror' to the back of the box.

Figure 8.1 (a) The law of reflection. (b) Image formation by a plane mirror.

When a ray of light is directed at a certain angle onto a flat mirror, the light ray reflects off the mirror at the same angle. In other words, the angle between the mirror and the light ray before reflection is the same as the angle between the mirror and the light ray after reflection. This statement is known as the **law of reflection** and it holds for any mirror. The law is usually expressed in the form

$$i = r$$

where i is the angle between the incident ray and the **normal** (the line at right angles to the mirror at the point of incidence) and r is the angle between the reflected ray and the normal.

Using this law, we can explain the distance rule by means of a ray diagram as shown in Figure 8.1(b). This diagram represents two light rays from a point object before and after reflection. The angle between each light ray and the mirror is the same before reflection as after reflection. Someone looking into the mirror along both reflected rays would see an image of the object at the position where the reflected rays appear to come from. The mirror in the ray diagram in Figure 8.1(b) is a line of symmetry. This is because each ray from the object to the mirror is at the same angle to the mirror as the corresponding 'apparent ray' from the image to the mirror. Therefore, the image and the object are the same distance from the mirror.

REFRACTION OF LIGHT

Refraction of light is the change of direction of a light ray when it passes from one transparent substance to another transparent substance. For example, when light passes from air into a lens, its direction changes, unless its initial direction was at right angles to the lens surface. Another example can be seen when the bottom of a swimming pool is viewed from above the water surface. The pool appears shallower than it really is. This is because light from a point on the bottom of the pool that passes through the surface non-normally is refracted away from the normal at the surface. As a result, the image of an object under water appears closer to the surface than the object really is.

Light is refracted

▶ *towards the normal when it passes from air into a transparent substance*
▶ *away from the normal when it passes from a transparent substance into air.*

Figure 8.2 (a) Refraction at a water surface. (b) Testing the law of refraction.

1   Observe a drinking straw in a cup of water. Notice that it appears to bend at the surface. This is because light refracts at the surface.

2   Look at an object through a magnifying glass. You should see an enlarged image of the object because light from the object passing through the magnifying glass is refracted. The image is further from the lens than the object.

3   Place a glass of water on top of newspaper print and observe the print through the water. The print viewed through the water appears to be higher up than the rest of the print. Again, this is because light refracts at the surface.

The **law of refraction** was discovered by Willebrord Snell in Holland in 1618. He measured the angle of refraction, r, between the normal and the refracted ray for different values of the angle of incidence i (i.e. the angle between the normal and the incident ray). He found that the ratio sin i/sin r is always the same for a given transparent substance. This ratio is referred to as the **refractive index** (symbol n) of the substance. Figure 8.2 (b) shows how to test the law of refraction.

$$\text{Refractive index } n = \frac{\sin i}{\sin r}$$

Using the wave theory of light described on p. 135, it can be shown that the refractive index of a substance is equal to the speed of light in air/speed of light in the substance.

$$\text{Refractive index } n = \frac{\text{speed of light in air}}{\text{speed of light in the substance}}$$

Maths kit

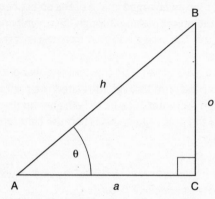

Figure 8.3 Trigonometry equations.

For the right-angled triangle ABCE shown in Figure 8.3, let the angle BAC be denoted by the Greek letter θ (pronounced θ). The following equations are used to define three trigonometry functions including sin θ (pronounced 'sine θ)', where o is the length of the side opposite to θ, h is the length of the side opposite the right angle and a is the length of the other side adjacent to θ:

$$\sin \theta = o/h$$
$$\cos \theta = a/h$$
$$\tan \theta = o/a$$

Worked example

A light ray enters a glass block at an angle of incidence of 30°
The angle of refraction was found to be 22° Calculate: (a) the
refractive index of the glass block, (b) the angle of refraction for
an angle of incidence of 60°.

SOLUTION

(a) $n = \sin 30/\sin 22 = 0.50/0.34 = 1.5$

(b) Rearrange $n = \sin i/\sin r$ *to give* $\sin r = (\sin i)/n = \sin 60/1.5 = 35°$

QUESTIONS

1 *If you stand 0.60 m in front of a flat mirror, how far away from
you is your own image?*

2 *State the law of reflection at a flat mirror.*

3 *Copy and complete the ray diagram in Figure 8.4, showing the
formation of an image by a flat mirror.*

Figure 8.4 Complete this diagram (see question 3).

4 (a) *What is meant by refraction of light?*
   (b) *A light ray is directed at a glass block of refractive index 1.5 at
       an angle of incidence of 60°, as shown in Figure 8.5. Calculate
       the angle of refraction of the light ray.*
   (c) *The speed of light in air is 300000 km/s. Show that the speed
       of light in glass of refractive index 1.5 is 200000 km/s.*

5 *Copy and complete the path of the light ray through the glass block in Figure 8.5.*

glass

light ray

Figure 8.5 See questions 4 and 5.

Theories of light

NEWTON'S THEORY OF LIGHT

If you have played or observed a game of snooker, you will know that a ball bounces off a wall at the same angle as it hits the wall – provided that it is not spinning as it rolls towards the wall. In addition to establishing laws on motion and gravity, Sir Isaac Newton put forward a theory of light in which he proposed that light consists of tiny particles which he called **corpuscles**. Perhaps Newton had the idea in his mind about a ball bouncing off a wall, for that was how he explained the reflection of light, supposing that the corpuscles of light bounced off a mirror in the way that a ball bounces off a wall. Newton needed to assume that the corpuscles did not lose any speed due to the impact so they moved away from the wall at the same angle as they had moved towards it.

Newton used his corpuscular theory to explain refraction of light as well as reflection. According to Newton, a corpuscle of light in air moving towards a transparent substance is attracted towards the substance. The force of attraction causes it to move faster in the substance than in air. More importantly, if its initial direction is not directly along the normal, the increase of speed causes its direction to change nearer to the normal.

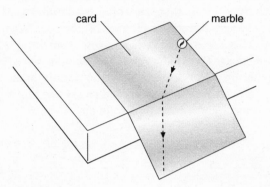

Figure 8.6 A refraction model.

A model of refraction can be made using a marble and a piece of card folded down the middle. One half of the card needs to be fixed to a book on a slight slope, with the other half of the card sloping steeply down to the surface on which the book rests. Observe the progress of a marble rolled slowly across the card on the book towards the fold. When it passes over the fold onto the steeper section, its speed increases and its direction changes at the fold.

HUYGENS' WAVE THEORY OF LIGHT

About the same time as Newton put forward the corpuscular theory of light, an alternative theory of light was put forward in Holland by Christiaan Huygens. According to Huygens, light consists of a wave motion which moves through space or air or any transparent substance in the way that waves move across a water surface. This imaginative theory also provided an explanation of the laws of reflection and refraction of light. Reflection of water waves is observed when sea waves bounce off a harbour wall. Refraction of water waves is the reason why waves on a beach usually run straight up the beach, regardless of the direction of the waves as they approach the beach. The waves slow down as the water becomes less deep and this reduction in speed causes them to move directly towards the beach. This situation is not unlike a front-wheel drive vehicle that veers off a road onto muddy ground. The vehicle changes direction as it goes off the road because one of the front wheels leaves the road and loses some of its grip before the other one does. To explain refraction of light, Huygens needed to assume that light travels more slowly in a transparent substance than in air. In contrast, Newton needed to assume that light travels faster in a transparent substance than in air.

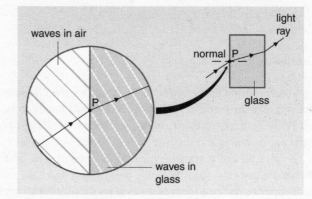

Figure 8.7 Using wave theory to explain refraction.

YOUNG'S DOUBLE-SLITS EXPERIMENT

Reflection and refraction of light can both be explained using either Newton's corpuscular theory or Huygens' wave theory of light. Newton's theory assumes that light travels faster in a transparent substance than in air, whereas Huygens' theory assumes that it travels more slowly in a transparent substance. Which theory is correct? When the theories were compared in the seventeenth century, most people believed Newton, since the speed of light could not be measured and Newton's scientific reputation was much greater than that of Huygens. Newton's theories of motion and gravity were outstandingly successful in explaining the motion of objects in all possible situations. So Newton's theory of light was accepted for over a century, even though there was no direct evidence for the corpuscular theory of light.

The idea that light consists of tiny particles held sway until the first decade of the nineteenth century, when Thomas Young at the Royal Institution in London used light to demonstrate the phenomenon of **interference**. Young showed that if a narrow source of light is observed through two closely spaced slits, a series of bright and dark bands (referred to as 'fringes') are seen. The fringes are replaced by a broad band of light if one of the two slits is blocked. Thus the dark bands are formed where light from one slit cancels out light from the other slit. The light from the two slits is said to interfere

▶ *constructively, where a bright fringe is formed – the light from each slit reinforces the light from the other slit*
▶ *destructively, where a dark fringe is formed – the light from each slit cancels the light from the other slit.*

Observing interference
Use the double-slit arrangement shown in Figure 8.8 to observe interference fringes using light from a torch lamp adapted as a narrow source. You ought to be able to see bright and dark parallel fringes.

Insight
The slits need to be narrow and less than about 0.5 mm apart. A pin could be used to make two closely-spaced slits in a strip of metal foil. The light source needs to be as narrow as possible. For example, a thin gap between pieces of insulation tape over the front of a torch lamp could be used.

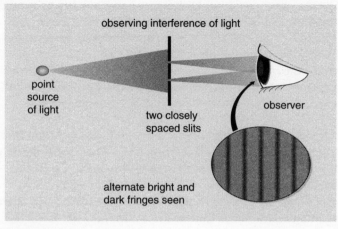

Figure 8.8 Observation of interference fringes.

The phenomenon of interference can also be demonstrated by sending straight waves on a water surface towards two narrow, closely spaced gaps in a barrier in the water. The waves that pass through the gaps spread out and overlap. In the overlap region, wave crests and troughs from one slit overlap and pass through crests and troughs from the other slit.

▶ *Reinforcement occurs where a crest meets a crest or a trough meets a trough. With light, this occurs at each bright fringe.*
▶ *Cancellation occurs where a crest meets a trough. With light, this occurs at each dark fringe.*

Young demonstrated the phenomenon of interference of light before an audience of invited guests at the Royal Institution. However, he was unable to convince them that light was wave-like in its nature. They preferred to believe that some unknown property of corpuscles would explain the phenonemon of interference. The dispute about whether light consists of waves or particles was only settled several decades later when it was shown

clearly that light travels more slowly in water than in air –
as predicted by Huygens over a century earlier!

The wavelength of light, the distance from one crest of a wave to
the next crest, depends on its colour. By making measurements
using Young's slits arrangement, the wavelength of each colour
can be determined. The wavelength of each colour is shown in
Figure 8.10. Note that the wavelength decreases from about
0.0007 millimetres for red light to about 0.0004 millimetres for
violet light. Two thousand wavelengths of yellow light would fit
into the space between adjacent millimetre marks on a ruler.

The smallness of the wavelength of light is the reason why wave
effects are difficult to demonstrate with light.

THE SPEED OF LIGHT

Light travels through space at a speed of 300 000 kilometres per
second. Light from the Sun takes about 500 seconds to reach the
Earth. Light from the most distant galaxies reaching us now was
emitted thousands of millions of years ago. The first accurate
measurement of the speed of light was made by Armand Fizeau
in France in 1849. He observed light from a narrow source of
light after it had travelled a distance of over 17 kilometres.
A narrow beam of light from the source was directed at the teeth
of a rotating cog wheel, which chopped the beam into pulses. The
beam was then directed to a distant mirror where it was reflected
back to the edge of the cog wheel. The beam passes back through
the edge of the cog wheel if a gap between the teeth is present
when each pulse returns to the wheel. Figure 8.9 shows the idea.
If the wheel is turned faster and faster from zero speed, an observer
looking at the image of the light source along the returning beam
would see the image at certain rotation speeds only. At each such
rotation speed, the time taken by the beam to travel from the wheel
and back again is equal to the time taken for a gap to be replaced
by another gap at the light beam.

Figure 8.9 Measuring the speed of light.

For a wheel with N teeth turning at a rotation frequency of f turns per second,

▶ *the time taken for 1 rotation of the wheel = 1/f,*
▶ *the time taken for one gap to replace the next gap,*
  t = time for 1 rotation/N = 1/fN.

If the light travels a distance D in this time (from the wheel and back), the speed of light, c = distance/time taken = $D/(1/fN)$ = DfN.

Fizeau used a wheel with 720 teeth and found that the lowest frequency at which the image reappeared was 25.2 rotations per second when the total distance from the wheel and back was 17.3 km. Prove for yourself that these measurements give a value for the speed of light of 314 000 kilometres per second. Fizeau used this method to show that light travels more slowly through water than through air. Improved methods for the measurement of the speed of light were subsequently devised. The speed of light in a vacuum is now defined as 299 792.458 kilometres per second.

ELECTROMAGNETIC WAVES

Water waves are disturbances that travel across the surface. Sound waves in air are pressure variations that travel through the air.

Seismic waves are vibrations that travel through the Earth after being created in an earthquake. What are light waves? In the nineteenth century it was generally thought that light consists of vibrations in an invisible substance referred to as 'ether', which was thought to fill space. In 1862 a mathematical theory of light waves was published by the Scottish physicist James Maxwell. He showed that light consists of vibrating electric and magnetic fields in which the electric vibrations generate magnetic vibrations which generate electric vibrations, etc. Such waves were referred to as **electromagnetic waves**. Maxwell combined the theory of electric fields with the theory of magnetic fields to show that such electromagnetic waves ought to travel through space at a speed of 300 000 kilometres per second. He knew that light travels at this speed through space, so he concluded that

▶ *light consists of electromagnetic waves, and*
▶ *electromagnetic waves exist beyond both ends of the visible spectrum.*

Figure 8.10 The electromagnetic spectrum.

When Maxwell published his theory of electromagnetic waves, it was immediately realized that infrared radiation, which had been detected beyond the red part of the visible spectrum, and ultraviolet radiation, which was known to lie beyond the violet part of the visible spectrum, must also consist of electromagnetic waves. Could there be electromagnetic waves even longer in wavelength than infrared radiation? What about beyond ultraviolet radiation? Several decades after Maxwell published his theory, the German physicist Heinrich Hertz discovered how to produce and detect electromagnetic waves much longer in wavelength than infrared radiation. These waves became known as radio waves and within a few years they were being used to transmit wireless signals between Britain and America.

TUNING IN

When you tune in to a radio or TV station, you are adjusting a receiver circuit so that it will accept radio or TV waves of a certain wavelength only. The **wavelength** of a wave is the distance along the wave from one wave crest to the next wave crest. The Greek symbol λ (pronounced 'lambda') is used for wavelength. The tuner of a radio receiver usually displays the wavelength or the frequency of the waves that the tuner accepts.

▶ *The* **frequency** *is the number of complete cycles of waves passing a point each second, where one cycle is from one wave crest to the next wave crest. The unit of frequency is the hertz (abbreviated as Hz), where 1 Hz = 1 cycle per second. Local radio stations broadcast at frequencies of about 100 megahertz (MHz), where 1 MHz = 1 million hertz.*

▶ *The* **speed** *of waves is the distance per second travelled by a wave crest. For waves moving at speed v, in 1 second a wave crest would travel a distance equal to v. The number of wavelengths in this distance is equal to v/λ, where λ is the wavelength of the waves. Hence the frequency f of the waves, which is the number of waves passing a given point in 1 second, is equal to v/λ.*

Frequency, f = speed of waves, v/wavelength, λ

Hence for electromagnetic waves of wavelength λ, the frequency of the waves $f = c/\lambda$, where c is the speed of electromagnetic waves. For transmission through air or through space, c = 300 000 kilometres per second.

Worked example

Calculate: (a) the frequency, in air, of radio waves of wavelength 100 metres, (b) the wavelength, in air, of radio waves of frequency 100 MHz. The speed of electromagnetic waves in air = 300000 kilometres per second.

SOLUTION

(a) Frequency f = speed c/wavelength λ = 300000 × 1000 metres per second/100 metres = 3 000 000 Hz = 3.0 MHz.

(b) Multiplying both sides of $f = c/\lambda$ by λ gives $f\lambda = c$.

Dividing both sides of $f\lambda = c$ by f gives $\lambda = c/f$.

Hence λ = 300000 × 1000 metres per second/100 000 000 Hz = 3.0 m.

	frequency	range uses
long wave (LW)	up to 300 kHz	international AM radio
medium wave (MW)	300 kHz–3 MHz	AM radio
high frequency (HF)	3–30 MHz	AM radio
very high frequency (VHF)	30–300 MHz	FM radio
ultra high frequency (UHF)	300–3000 MHZ	TV broadcasting, mobile phones
microwave	above 3000 MHz	satellite TV, global phone links
light	500 THz approx	fibre optic communication links

Note: 1 MHz = 1000000 Hz. 1 THz = 1 million MHz.

Figure 8.11 Frequency bands.

Insight

In many countries television channels are being switched, region by region, to digital channels at higher frequencies. The changeover will allow more channels for wireless broadband and high-definition TV (HDTV).

A few years after the discovery of radio waves, the German physicist Wilhelm Röntgen discovered how to produce and detect electromagnetic waves much shorter than ultraviolet radiation. These waves became known as **X-rays**. Their discovery received great publicity as newspapers claimed they could be used to 'see' through objects. In fact, X-ray machines were quickly put to use in hospitals to photograph broken bones in limbs. This is possible because X-rays are absorbed by dense materials such as bone and pass through soft material such as human tissue. If a beam of X-rays is directed at photographic film in a light-proof wrapper, the film is blackened where X-rays reach it, because the X-rays, unlike light, can pass through the wrapper. If a limb is placed in the path of the beam before it reaches the film, a shadow of the bone in the limb can be seen on the film when it is developed, because the bone prevents X-rays from reaching the film.

Insight

X-rays are harmful. Like the radiation from radioactive substances, they knock electrons out of atoms, causing the atoms to become charged. Charged atoms are called ions. Their creation by ionizing radiation in a body cell can damage or kill the cell.

In the same decade as X-rays were discovered, the discovery of radioactivity by Henri Becquerel in France led to the conclusion that radioactive substances can emit electromagnetic waves even shorter than X-rays. This type of radiation is known as **gamma radiation** (or γ radiation using the Greek letter γ, pronounced 'gamma'). Maxwell's theory of electromagnetic waves was thus confirmed by these discoveries. Predictions from the theory were used to develop many practical applications, such as the design of radio transmitters on ships and of X-ray tubes for use in hospitals.

Figure 8.12 X-rays.

3 The kinetic energy of each electron is converted into heat and X-radiation. A beam of X-rays spreads out from the spot on the anode where the electrons strike it. The spot becomes very hot as most of the kinetic energy of each electron is converted to heat.

anode

2 These electrons are attracted to the metal **anode** which is at a high positive potential. They are accelerated to high speeds and then stopped by collision with the anode. The glass tube is evacuated so that electrons can reach the anode from the filament.

1 The **filament** wire is heated by passing an electric current through it. This causes it to emit electrons.

4 Thick lead shields surround the tube to ensure X-rays do not emerge in unwanted directions. Two sets of thicker lead plates are used to restrict the beam to the part of the patient under treatment.

5 X-rays are absorbed by bones and pass through soft tissues to form 'shadow' image of the patient's bones on photographic film.

Classical physics based on Newton's laws and Maxwell's theory of electromagnetic waves was able to account for every known phenomenon or observation. Many physicists towards the end of the nineteenth century thought that the laws of nature had more or less been discovered, using the theories of classical physics. It seemed that little remained to be explained, although more accurate measurements of the properties of materials and light would probably justify the scientists' continued efforts. Some niggling minor discoveries were proving troublesome to explain, but by the turn of the century many physicists were satisfied that the laws of nature were mostly known.

PHOTOELECTRICITY AND PHOTONS

When Hertz was investigating how to produce and detect radio waves, he noticed that the sparks induced by radio waves in a 'spark-gap' detector were stronger when ultraviolet radiation was directed at the spark gap contacts. Hertz was more interested in radio waves and so he passed his observation on to other physicists to investigate. Further investigations showed that the effect happens because a metal emits tiny negative particles identified as electrons when illuminated with light of frequency above a certain value. This happens when ultraviolet radiation is directed at a metal because ultraviolet has a much higher frequency than any colour of light. The investigators knew that electrons are contained in every atom. They also knew that metals conduct electricity because some of the electrons in a metal move about inside the metal, not confined to individual atoms. Now they had found that light could be used to make some of these electrons escape from the metal. This effect is called the **photoelectric effect**.

Precise observations using light of different frequencies produced the astonishing conclusion that the frequency of the light needed to be greater than or equal to a certain 'threshold' value that depended on the metal being tested. The existence of such a threshold frequency could not be explained using the wave theory

light

metal
plate

electron

Figure 8.13 Photoelectricity.

of light. According to wave theory, the effect should happen
regardless of the frequency of the light. It was expected that the
effect would happen more slowly, the lower the frequency of light
used – but nevertheless it would happen, according to wave theory.
However, the experimental results showed that electrons were not
emitted at all if the frequency of the light was below the threshold
frequency. In 1905, this troublesome problem from the last few
years of the nineteenth century was to shatter the well-established
classical wave theory of light.

EINSTEIN'S PHOTON THEORY OF LIGHT

Albert Einstein was an awkward student at the University in
Zurich, for he kept asking questions that his professors were
unwilling or unable to answer. He graduated with a degree in
physics in 1902 and eventually secured a post as a patent officer
in Berne, Switzerland. He attended to his official duties efficiently
and was able to pursue his ideas about physics in his spare time.
In 1905 he wrote several scientific papers which revolutionized
physics and recast the laws of physics. We shall return to his
mind-boggling ideas about space and time in Chapter 10.
For the moment we will concentrate on his explanation of
photoelectricity, for which he was awarded the Nobel Prize
for physics in 1921.

Einstein explained photoelectricity in his 1905 paper on the interaction of light and matter by inventing a new theory of light which he called the **photon theory of light**. Einstein assumed that light is composed of wave packets which he called photons. The key points of photon theory are:

1 *Each photon is a packet of electromagnetic waves moving in a particular direction, not spreading in all directions as in classical wave theory.*

2 *The energy of a photon is in proportion to the frequency of the waves in the wave packet. Einstein used an earlier idea about energy from Max Planck in Berlin, which we shall meet in Chapter 10, and assumed that the energy E of a photon is given by the equation*

$$E = hf$$

where h is a constant referred to as the Planck constant.

To explain photoelectricity using photon theory, Einstein said that:

1 *An electron in a metal needs a minimum amount of energy to escape. This amount of energy is called the work function of the metal.*

2 *An electron near the surface can escape if it gains energy equal to or greater than the work function of the metal.*

3 *An electron is able to escape from the surface if it absorbs a single photon of energy greater than the work function of the metal.*

Thus if an electron near the surface absorbs a photon of energy hf, it can escape if hf is greater than or equal to W, the work function of the metal. The threshold frequency of the light therefore is equal to W/h, corresponding to $hf = W$.

Hence, for photoelectric emission to occur, the frequency of the incident light f must be greater than or equal to the threshold frequency. Thus Einstein's photon theory provides an explanation of photoelectricity. Further predictions about the interaction of light and matter were confirmed by more investigations after 1905, and by 1921 the photon theory had been fully accepted.

Insight

Einstein imagined photons as 'flying needles'. Light photons are obviously harmless – but ultraviolet and X-ray photons aren't.

THE STORY OF LIGHT SO FAR

In the seventeenth century Newton thought that light was corpuscular. Huygens' wave theory of light was not accepted until the early decades of the nineteenth century when light was shown to travel more slowly in water than in air, as predicted by Huygens but not by Newton. Interference of light was discovered by Young in the first decade of the nineteenth century but was not accepted at that time as evidence for the wave theory of light. The particle nature of light was reintroduced as a result of the discovery of photoelectricity and its subsequent explanation by Einstein in terms of his photon theory. Light has a dual nature: it is wave-like in interference experiments and particle-like in photoelectricity. This dual nature of light was to provide a path in a direction completely unexpected by physicists at the end of the nineteenth century. Within two decades, the classical laws of physics had given way as a revolution swept through physics and our perception of matter and energy. We will meet the new physics, known as quantum physics, in Chapter 10.

Electromagnetic waves

1   The spectrum of electromagnetic waves in order of increasing wavelength: gamma rays and X-rays; ultraviolet radiation; visible light; infrared light; microwaves; radio waves.

(Contd)

Photon theory:

1   *Light consists of photons which are wave packets of
    electromagnetic radiation. The energy of a photon $E = hf$,
    where h is the Planck constant and f is the frequency of the
    light.*
2   *Photoelectricity is explained by the photon theory by assuming
    that the energy gained by a conduction electron in a metal
    which absorbs a photon (= hf) is sufficient to enable the
    electron to leave the surface of the metal if hf is more than the
    work function of the metal.*

QUESTIONS

6   (a) *What was the nature of light according to Newton?*
    (b) *Why was the wave theory of light eventually accepted?*

7   *Explain why a pattern of bright and dark fringes is seen when light
    from a narrow source is observed through a pair of closely spaced
    double slits.*

8   (a) *What is a photon of light?*
    (b) *What is the photoelectric effect?*
    (c) *What aspect of the photoelectric effect could not be explained
        by wave theory?*
    (d) *How does the photon theory explain the photoelectric effect?*

9   (a) *List the following parts of the electromagnetic spectrum in
        order of increasing wavelength:
        gamma rays, infrared radiation, microwaves, radio waves,
        ultraviolet radiation, visible light.*

(b) What types of radiation are beyond the long wavelength end of the visible spectrum?

10 Light from a laser beam has a wavelength of 0.0006 mm.
   (a) What is the colour of this light?
   (b) Calculate the frequency of light of this wavelength. The speed of light is 300 000 kilometres per second.

10 THINGS TO REMEMBER

1 *For a light ray that reflects from a plane mirror, its angle of incidence i = its angle of reflection r.*

2 *Light travels at a speed of 300 000 km/s in space and in air.*

3 *The* **refractive index** *of a transparent substance n = the speed of light in air/the speed of light in the substance.*

4 *A light ray passing from air into a transparent substance at an angle of incidence i refracts towards the normal such that sin i/sin r = n, where r is the angle of refraction.*

5 *Newton's* **corpuscular theory** *of light predicted that light travels faster in glass than in air; Huygens'* **wave theory** *predicted correctly that light travels more slowly in glass than in air.*

6 **Interference** *fringes can be seen when light from a point source passes through two closely spaced slits.*

7 *Light is part of the spectrum of* **electromagnetic waves.**

8 *The* **photoelectric effect** *is the emission of electrons from a metal plate when light above a certain frequency, dependent on the metal, is directed at the plate.*

9 *Einstein put forward the* **photon theory of light** *to explain the photoelectric effect as the wave theory of light could not explain the effect.*

10 *A photon is a packet of electromagnetic waves of energy E = hf, where f is the light frequency and h is the Planck constant.*

9

Materials and molecules

In this chapter you will learn:
- *about atoms and what is inside them*
- *how atoms join together to form molecules*
- *why solids, liquids and gases have characteristic properties.*

Many new materials have been discovered or invented since 1900 by scientists extending their knowledge about existing materials. Before the Scientific Age most materials were natural products or made from minerals. The Scientific Age brought new processes, new products and new discoveries, which have been put to use to make our lives better. A list of materials used by each of us every day would be very long. So it's all the more surprising that all materials and substances, manufactured or natural, are made from no more than 92 basic substances known as chemical elements. Before the Scientific Age all substances were thought to be made from four basic elements, namely earth, water, air and fire. Progress to our present knowledge is due to the efforts of many scientists whose experiments and observations gave rise to scientific theories to explain the experimental data and then to further predictions and more experiments. New materials continue to be discovered in the same way, often leading to new devices and products. Understanding the properties of any substance requires underpinning knowledge and a sound grasp of certain key ideas. In this chapter we will look at the development of these ideas and see how they provide the foundations of the science of materials.

Atoms and molecules

How would you describe the differences between solids, liquids and gases? Solids have their own shape whereas liquids and gases do not. Liquids and gases can flow whereas solids cannot. Liquids and solids have surfaces whereas gases do not. The three states of matter, solid, liquid and gas, are described as physical states because a substance can be changed from one state into another and back into its former state by heating and cooling. For example, ice melts and becomes water if heated. In contrast, chemical changes involve reactions between substances that cannot be reversed by simple physical means such as heating and cooling. For example, water and carbon dioxide are formed when a candle burns, but candle wax cannot easily be formed from water and carbon dioxide. The obvious differences between the three states of matter are explained using two key ideas:

▶ *All substances consist of* **molecules***. A molecule is the smallest particle of a substance that can exist independently. The molecules of a pure substance are identical to one another and differ from the molecules of any other substance.*

▶ *Molecules link together in the liquid state and the solid state because they exert forces on each other. These forces are referred to as* **bonds***, because they bind the substance in the liquid or the solid state. Energy supplied to a solid at its melting point causes it to melt because the molecules gain sufficient energy to break free from each other. Energy supplied to a liquid at its boiling point causes it to vaporize because the molecules gain sufficient energy to move away from each other.*

COMPOUNDS AND ELEMENTS

A **compound** is a pure substance which consists of one type of molecule only. A compound can be broken down into other substances.

An **element** is a pure substance that cannot be broken down into other substances. There are just 92 naturally occurring elements. Further short-lived elements have been discovered in nuclear reactions, but none of these is found naturally.

An **atom** is the smallest part of an element that is characteristic of the element. The atoms of an element are identical to one another and differ from the atoms of any other element. The lightest atom is the hydrogen atom. The heaviest naturally occurring atom is the uranium atom, which is 238 times heavier than the hydrogen atom.

A **molecule** consists of two or more atoms joined together by force bonds. Each type of molecule consists of a fixed number and type of atoms. For example, every carbon dioxide molecule consists of one carbon atom and two oxygen atoms. Its chemical formula is therefore written as CO_2.

The idea of atoms was first put forward in Ancient Greece by the philosopher Democritus (470–400 BCE), who considered objects to be made of tiny indivisible and indestructible particles he called atoms. His theory was rejected by Aristotle, in favour of the theory of 'the elements', earth, water, air and fire. The atomic theory was established in its modern form in the early nineteenth century by John Dalton (1766–1844). He knew from the work of other scientists that the composition by weight of the elements in a given compound is always the same. He realized that this is because a compound consists of compound atoms or molecules, each molecule being composed of a fixed number of atoms of the elements that the compound can be broken down into. He worked out the relative weight of each type of atom, based on one unit of atomic weight for the hydrogen atom. He drew up a table of atomic weights for all the known elements and used it to explain why elements combine in simple proportions when they form compounds. Dalton's atomic theory led to the Periodic Table, which is the basis of modern chemistry and which can only be explained using quantum mechanics (see p. 180).

INSIDE AN ATOM

By the end of the nineteenth century scientists knew that atoms are not indivisible or indestructible. Experiments on electrical discharges through gases at very low pressures resulted in the discovery of electrons. They were pulled out of the gas atoms by the strong electrical field applied to the discharge tube. Regardless of the type of gas used, it was discovered that identical negatively charged particles were produced as well as positively charged atoms of differing weights. It was therefore concluded that these negatively charged particles, referred to as electrons, are in every type of atom. Further investigations on radioactivity (see Chapter 11) led to the conclusions that:

▶ *Every atom contains a positively charged nucleus where most of its mass is concentrated.*
▶ *The nucleus is composed of two types of particle, protons and neutrons. The nucleus of the hydrogen atom is a single* **proton***.*

▶ The proton is a positively charged particle and is slightly lighter than the **neutron**, *which is uncharged.*

▶ Electrons move in the space round the nucleus at relatively large distances. The charge of the electron is equal and opposite to the charge of the proton. The electron is about 2000 times lighter than the proton.

	charge/ proton charge	mass/ proton mass
proton	1	1
neutron	0	1
electron	−1	0

Figure 9.1 The structure of the atom.

Insight

The diameter of the nucleus in relation to the diameter of the atom is about the same as that of a small ball at the centre of a large stadium. Learn more about the nucleus in Chapter 11.

An atom is represented by the symbol $^A_Z X$ where Z is the proton number of the atom and A, the mass number, is the number of protons and neutrons in the nucleus. The mass of an atom is approximately equal to A times the mass of a hydrogen atom. This is because A is the number of protons and neutrons in its nucleus and each proton or neutron has a mass approximately equal to the mass of a hydrogen atom. An uncharged atom has the same number of electrons and protons. For example, the symbol $^7_3 Li$ represents an atom of lithium with three protons and four neutrons in its nucleus and three electrons moving about the nucleus. The mass of such an atom is approximately seven times the mass of a hydrogen atom.

ISOTOPES

Each atom of the same element contains the same number of protons in its nucleus. The number of neutrons in each atom of an element can differ. Atoms of an element with different numbers of neutrons are referred to as **isotopes**. For example, natural uranium mostly consists of the isotope $^{238}_{92}U$ and a small proportion of the isotope $^{235}_{92}U$. Both types of atoms are uranium atoms, each nucleus containing 92 protons. However, the isotope $^{238}_{92}U$ contains three more neutrons than the isotope $^{235}_{92}U$.

Note that the number of electrons in an uncharged atom is equal to the number of protons in the nucleus. The chemical properties of an element are the same for all the isotopes of the element. This is because chemical reactions are determined by the electrons in an atom. Atoms of the same element undergo the same chemical reactions, because each atom has the same electron arrangement even if the atoms are different isotopes of the same element.

The atomic scale of mass

The mass of an atom or a molecule is usually expressed in atomic mass units (symbol u) where $1\ u = 1.66 \times 10^{-27}$ kilograms. One unit of atomic mass (1 u) on this scale is defined as 1/12 of the mass of a $^{12}_6C$ atom. This type of atom is used because it can easily be isolated from other carbon atoms.

Insight

The unit of atomic mass was originally defined as the mass of a hydrogen atom. This was changed after it was discovered that hydrogen gas contains a tiny proportion of 'heavy hydrogen' atoms – atoms with a nucleus containing a neutron (in some instances two neutrons) as well as a proton.

QUESTIONS

1 (a) What type of charge is carried by (i) an electron, (ii) a proton?
  (b) The most common isotope of oxygen has 8 electrons, 8 protons and 8 neutrons in each atom.

(i) What is the mass of this type of atom in atomic mass units?

(ii) What is the total charge of the nucleus of this type of atom?

(iii) What would be the overall charge of such an atom if it lost one electron?

2   Use the following words to fill the gaps in the sentences below.

atom   electron   isotope   molecule   neutron   proton.

(a) Each _____ of carbon dioxide CO_2 consists of two oxygen _____ and one carbon _____ .

(b) $^{238}_{92}U$ is an _____ of uranium. Each nucleus of $^{238}_{92}U$ is composed of 92 _____ and 146 _____.

(c) The lightest atom is the hydrogen isotope $^{1}_{1}H$. This type of atom consists of one _____ and one _____ .

_____

More about bonds

The atoms in a molecule are joined to each other by **bonds,** which hold the atoms together. Atoms and molecules in liquids and solids form bonds that prevent the particles moving away from each other. If the bonds are sufficiently strong, the molecules lock each other together in a rigid structure. There are several different types of bond. The electron arrangement of each type of atom determines the type of bond formed and this determines the physical state of the substance at a given temperature.

The electrons in an atom move round the nucleus like planets moving round the Sun. The electrical attraction between each electron and the nucleus prevents the electron from leaving the atom. The electrons in an atom can only occupy certain orbits round the nucleus. These allowed orbits are called **shells.** Each shell can hold up to a certain number of electrons. The innermost shell of an atom can only hold two electrons. The next shell from the nucleus can hold up to eight electrons. The nearer a shell is to the nucleus, the lower the energy of an electron in the shell.

The electrons in an atom fill the shells from the innermost shell outwards. Each electron in a shell can escape to an outer shell only if it is given a certain amount of energy. The table below shows the arrangement of the electrons in the 11 lightest atoms. These occupancy rules were worked out from the chemical properties of the elements and can be explained using **quantum mechanics**, which is introduced in the next chapter.

| Element | Atomic number Z | Number of electrons in each shell | | |
		1st shell	2nd shell	3rd shell
Hydrogen	1	1	0	0
Helium	2	2	0	0
Lithium	3	2	1	0
Beryllium	4	2	2	0
Boron	5	2	3	0
Carbon	6	2	4	0
Nitrogen	7	2	5	0
Oxygen	8	2	6	0
Fluorine	9	2	7	0
Neon	10	2	8	0
Sodium	11	2	8	1

The chemical properties of an element depend on the electron arrangement of its atoms. Because a completely empty or full shell is in a lower energy state than a partially full shell, atoms react by gaining or losing electrons to make full shells only. Thus an atom of sodium reacts by losing the electron in its third shell, whereas an atom of fluorine reacts by accepting an extra electron into its second shell. Other types of atom react by sharing electrons to form completely full shells.

The **Periodic Table** is a table of all the known elements listed in rows in order of increasing atomic number. The elements in each column of the Periodic Table are referred to as a 'group', as they have similar chemical properties. This is because each type of atom in a column has the same number of electrons in a partially filled or completely filled outer shell. For example, lithium, sodium, potassium and certain heavier elements, all in column 1, are metals that readily react because each type of atom in this group has a single electron in the outer shell. This lone electron is easily removed when the atom reacts with another atom. Another distinctive group of elements are the inert gases: helium, neon, argon, xenon and krypton. Atoms of all these elements have their electrons in completely filled shells and are therefore unreactive.

TYPES OF BOND

The type of bond formed between two atoms or molecules depends on how the atoms or molecules can lose or gain or share electrons to reach a lower energy state.

▶ *A **covalent bond** is formed as a result of two atoms sharing a pair of electrons. Each atom contributes an electron to the bond from its outer shell. For example, hydrogen gas consists of molecules which each consist of two hydrogen atoms. The two atoms form a covalent bond by sharing their two electrons so each of the atoms in the molecule has a full shell of electrons. Another example is provided by the carbon dioxide molecule, which consists of one carbon atom and two oxygen atoms. Each oxygen atom forms two covalent bonds with the carbon atom. The second shell of the carbon atom in a molecule is therefore filled with eight electrons, four originally belonging to the carbon atom and two from each oxygen atom. The second shell of each oxygen atom in the molecule also has eight electrons, six from the oxygen atom and two from the carbon atom.*

Figure 9.2 Covalent bonds.

+ Sodium ion

− Chloride ion

Each ion is attracted
equally to its 4 nearest
neighbours so is held in
its place

Figure 9.3 Ionic bonding.

▶ *An ionic bond is formed when an atom of one element loses
an electron in its outer shell to an atom of a different element.
The donor atom becomes positively charged because it loses
an electron, whereas the atom that gains the electron becomes
negatively charged. Charged atoms are called ions. Hence
the atoms that form ionic bonds attract each other because
they are oppositely charged. The characteristic shape of a
crystal occurs because it contains positive and negative ions
arranged in a regular pattern referred to as a lattice. Each
ion is held in place in the lattice by ionic bonds formed with
adjacent, oppositely charged ions. Ionic crystals dissolve in
water because water molecules weaken the forces of attraction
between the ions.*

Crystal formation

Observe a few grains of common salt under a magnifying glass
and you will see that the grains are cubic in shape. Watch the
grains dissolve at the edge of a drop of water. Then blow steadily
over the water to make it evaporate. You will see the grains form
again. The chemical name of common salt is sodium chloride.
It contains positively charged sodium ions and negatively
charged chloride ions.

Insight

Common salt is a compound of sodium and chlorine. Each atom of chlorine gains an electron from an atom of sodium. The chlorine atoms therefore become negative ions because they each gain an electron. The atoms of sodium become positive ions because they each lose an electron. The ions form a cubic lattice.

▶ **Metallic bonds** *are formed in a solid metal, which consists of a lattice of positive metal ions surrounded by electrons that move about freely inside the metal. Each metal atom has lost one or more of its outer electrons. These electrons are referred to as* **conduction electrons** *because they carry electric charge through the metal when a voltage is placed across it. The conduction electrons prevent the positive ions from moving out of place in the lattice. Solid metals are much stronger than many other solids because metallic bonds are equally strong in all directions.*

▶ **Molecular bonds** *act between uncharged molecules at close range. This type of bond is much weaker than the other types of bond described above. A molecular bond is formed when two molecules are so close that the electrons of each molecule are slightly attracted to the nucleus of the other molecule. Molecules in liquids move about at random, but remain in the liquid because the bonds between the molecules prevent them from leaving the liquid surface. If the liquid temperature is raised, the faster-moving molecules near the surface can break away from the liquid and become gas molecules.*

QUESTIONS

3 *Explain in terms of electrons why:*
   (a) helium atoms do not interact with other types of atom
   (b) sodium ions carry a positive charge.

4 *What type of bond is formed when:*
   (a) two atoms share a pair of electrons?
   (b) one type of atom gains an electron and one type loses an electron?

Solids and structure

A solid object has its own natural shape because the force bonds lock the atoms of the solid together. When a solid object is stretched or compressed or twisted, the atoms are pulled away from each other. If the forces distorting an object are removed and the object returns to its natural shape, the object is said to possess **elasticity**. If the distorting forces are sufficiently large, the object will not regain its natural shape and is permanently distorted. For example, a Perspex ruler that is bent slightly becomes straight again when the bending forces are removed. If the bending forces are large enough, the ruler bends permanently or cracks. Another example is a paper clip, which is designed to be strong enough to hold sheets of paper together but not too strong or it would be impossible to use. If too much force is applied to a paper clip, it bends permanently. The limit beyond which an object loses its elasticity is referred to as its **elastic limit**. Above its elastic limit, the object is permanently distorted and is said to be **plastic** rather than elastic. The bonds between atoms in the object break and are re-formed with different atoms when the object shows plastic behaviour.

Elasticity tests

Carry out each of the following tests and decide if the elastic limit has been exceeded in each case.

1 Measure the length of a rubber band using a millimetre ruler. Stretch the rubber band and then release it. Measure its unstretched length again.
2 Repeat the above test with a strip of polythene cut from a polythene bag.
3 Repeat the test with a length of string.

You should find that the rubber band and the length of string remain within their own elastic limits but the polythene does not.

ATOMS IN SOLIDS

Solids may be classified as crystalline, amorphous or polymers.

1 Crystalline solids

The atoms in a crystal are arranged in a regular pattern. As a result, the crystal has a recognizable shape. For example, sodium chloride crystals are cubic because sodium and chloride ions occupy alternate positions – like opposing sports fans next to each other in rows of seats. Imagine each 'reds' fan sitting with an opposing 'blues' fan immediately in front, behind and on either side. The analogy ends there, as the sodium ions are much smaller than the chloride ions and carry opposite charge, which is why the structure holds together. Crystals with a particular shape consist of rows of atoms arranged in such a way as to create the characteristic shape of the crystal.

A crystal is difficult to break, but a small amount of force applied in a suitable direction can cause it to cleave into two smaller crystals. Such a force can be applied by tapping the crystal with a sharp edge parallel to one of its faces. The same amount of force applied differently to the crystal would have no effect on it. The reason why the crystal cleaves as above is because the tapping force is parallel to the layers of atoms in the crystal and it causes the atoms in one layer to slide over the atoms in an adjacent layer. The effect is not unlike applying a force to a stack of cards on a table so that the cards slide over each other. The same force pushing down on the stack of cards would have little effect on the stack.

Metals consist of tiny crystals called **grains**, packed together with no spaces between adjacent grains. To visualize the grain structure inside a metal, imagine an aerial view of a landscape consisting of small ploughed fields, with no spaces between them and a different direction of furrows in each one. The atoms in a grain are held together by metallic bonds but they arrange themselves in rows, like the furrows in a field. Across the boundary between two adjacent grains the rows change from one direction to a different direction, like the furrows in two adjacent fields. The grains in a metal can be observed using a powerful microscope, if the metal surface is cleaned and polished to make it very smooth. The strength of a metal is due to the presence of grain boundaries. When forces are applied to a metal within its elastic limit, the layers of atoms in each grain are unable to slide past each other because the layers in adjacent grains are not in the same direction. Imagine many packs of cards, each held by a rubber band, jumbled together in a large cardboard box. The box would be difficult to distort because the cards that could be made to slide by a sideways push force would be prevented from doing so by adjacent packs. If all the cards in all the packs were horizontal, then the box could easily be distorted by pushing it sideways.

TESTING STEEL

Heat treatment alters the strength of steel because it changes the size of the grains. If the end of a steel wire is heated in a flame and then suddenly cooled in water, the wire becomes very brittle and snaps easily. Heating the wire causes the grains to become larger and fewer and so there are fewer boundaries in the metal. Sudden quenching in water makes the enlarged grains permanent and the steel is therefore weaker, as there are fewer grain boundaries to stop layers of atoms sliding past each other when forces are applied to it.

2 Amorphous solids

An amorphous object has no characteristic shape. The atoms in an amorphous solid are locked together in contact with each other at random. In a liquid, the atoms are in contact with each other but they are not locked together and they move about in random directions. An amorphous solid is sometimes described as a 'frozen liquid', as if the atoms of a liquid suddenly stopped moving and locked together haphazardly. Glass is an example of an amorphous solid, consisting of silicon atoms and oxygen atoms linked together by covalent bonds. Glass is brittle because the effects of stress in the glass due to bending forces concentrate at tiny cracks in the surface. These cracks then deepen, which causes the stress in the glass to become even more concentrated so the glass snaps suddenly.

If glass is heated strongly it loses its brittleness, as cracks in the surface disappear, and it softens enough to be blown or drawn into any desired shape by a skilled glassblower. On cooling, it sets in its new shape and recovers its stiffness and brittleness.

Insight

Glassblowers are skilled at making distinctive glass objects. Glass is a poor conductor of heat so when part of it is heated, the heated part becomes very hot and soon softens.

3 Polymers

A polymer consists of long molecules, each consisting of a chain of atoms, formed as a result of identical shorter molecules joining together end-on. For example, polythene consists of long polyethylene molecules formed by making ethylene molecules join end-on. Each ethylene molecule consists of two carbon atoms and four hydrogen atoms. The two carbon atoms are joined together by a double bond (i.e. two covalent bonds) and two hydrogen atoms are attached to each carbon atom. When the ethylene molecules join end-to-end, the double bond is replaced by a single bond which enables the carbon atoms to link together in a long row. Each carbon atom in the polymer molecule is thus joined by a single covalent bond to a carbon atom on either side and by a covalent bond to each of two hydrogen atoms.

Many different polymers exist, all consisting of long molecules which are chains of identical shorter molecules. The long molecules of a polymer are tangled together in amorphous polymers such as polycarbonate, which is used to make 'hard hats'. The atoms in each chain are linked together by covalent bonds. Bonds form between chains where two chains are in contact. These cross-links can be strong (e.g. covalent bonds) or weak (e.g. molecular bonds), depending on the type of polymer. **Thermoplastics**, such as polycarbonate and amorphous polythene, are polymers with weak cross-links that break when the polymer is warmed, causing it to soften. These cross-links re-form when the polymer is cooled again and so the polymer can be moulded into any shape by warming then cooling it. **Thermosetting polymers**, such as Bakelite which is used for electrical fittings, have strong cross-links that make them stiff and heat-resistant.

Rubber is a natural polymer which consists of long molecules that tend to curl up. Cross-links between the molecules break when the rubber is stretched, and re-form when it is released. Raw rubber is made much stiffer by adding sulphur, in a process known as 'vulcanizing'. The sulphur atoms form strong cross-links between the rubber molecules, thus making the rubber stronger and stiffer.

A tyre tale

With every turn of a tyre, every part of the tread is squashed and then stretched as it makes contact with the road surface. Tyres are made of rubber because rubber is resilient, which means that it can be stretched and squashed repeatedly without losing its strength. Energy is used to stretch rubber and some of this energy is recovered when the rubber 'rebounds' to its normal shape. The unrecovered energy heats the rubber and raises its temperature. Fuel usage could be reduced if tyres were much stiffer, as less energy would be used to stretch and squash the tyres and so less energy would be lost as heat. However, very stiff tyres would have less contact with the road and would therefore be less safe. They would also make journeys very uncomfortable.

QUESTIONS

5 What is meant by (a) elasticity, (b) the elastic limit of an object?

6 State whether each of the following solid materials is crystalline, polymeric or amorphous: (a) steel, (b) polycarbonate, (c) sodium chloride, (d) glass, (e) rubber.

7 PVC is a polymer used to make electrical wires and cables.
   (a) What properties of PVC make it suitable for this use?
   (b) Why could it be dangerous if a PVC cable overheats?

Molecules in fluids

Any substance that can flow is a fluid. The molecules in a fluid are not locked together and they move about at random. In a gas, the molecules move very fast and are separated by large empty spaces. The molecules in a liquid move about more slowly than in a gas and they are in contact with each other. Liquids (and solids) are much denser than gases because the atoms and molecules in liquids (and solids) are much closer to each other than they are in gases. The forces between gas molecules are too small to affect the fast-moving molecules in a gas. However, in liquids and solids, the molecules move more slowly than in a gas and the forces between them are strong enough to pull the molecules together.

DIFFUSION

Observe a few drops of milk spreading out in a glass of water without stirring the water. The milk gradually spreads throughout the water. This process of the spreading of a fluid in another fluid is known as **diffusion**. It occurs because the molecules of each fluid are in motion and so they move about until they are evenly distributed. Diffusion takes place with gases as well as with liquids. For example, vapour from a liquid with a strong odour spreads out through the surrounding air. Someone near the liquid would notice the odour before someone further away. Initially, the vapour molecules are concentrated near the surface of the liquid but gradually they move away from the liquid in every direction until they are evenly distributed throughout the air. This process is the same as what happens when a container is filled with a layer of red beads covered with a layer of blue beads. If the lid is then closed and the container is shaken repeatedly, the beads gradually move about until they are distributed at random throughout the container. Provided there are a large number of beads, they become evenly distributed.

Diffusion tests
Some further examples of diffusion can easily be observed in the following situations.

1 *Release a drop of ink in a glass of water and observe the ink gradually spreading out. Eventually, the ink becomes evenly spread through the water.*

2 *Spray a small quantity of deodorant or perfume onto a cloth, then walk to the other side of the room. You should be able to detect the deodorant, using your nose, within a few minutes.*

3 *Pour a small quantity of vinegar onto a saucer and you should be able to detect its presence almost immediately, using your nose.*

Diffusion always results in an even spread of each substance throughout the container. This happens because the molecules move about at random and there is a huge number of molecules per cubic millimetre in any liquid or gas. If there were just a few molecules per cubic millimetre on average, the distribution of molecules would be uneven and constantly changing. However, with a large number of molecules per cubic millimetre, fluctuations in the distribution would be negligible. Imagine a thousand million molecules in a container of volume 1000 cubic millimetres (= 1 cubic centimetre). Each cubic millimetre would therefore contain 1 million molecules on average at any given instant. The probability of there being significantly more or less molecules in a cubic millimetre at any instant would be negligible. Diffusion results in even spreading because the possibility of uneven spreading is negligible for large numbers of molecules.

VISCOSITY

Water flows more easily than cooking oil, which flows more easily than syrup. The **viscosity** of a fluid is a measure of its resistance to flow. Syrup has a higher viscosity than cooking oil, which has a higher viscosity than water. In general, gases flow much more easily than liquids, so the viscosity of a gas is much lower than the viscosity of a liquid.

Viscosity depends on temperature. Oil flows more easily, the warmer it is. The lower the temperature of the oil, the less easily it flows. This is part of the reason why starting a vehicle can be difficult on a cold morning. The oil in the engine is cold and therefore less effective as a lubricant. Some liquids become more or less viscous when stirred. For example, paint in a tin becomes 'thinner' if it is stirred enough, which means that it flows more easily and is easier to apply. The opposite effect happens when wallpaper paste is stirred. Gradually it becomes 'thicker' and harder to stir until it reaches a consistency that allows it to be spread in a sticky paste onto the wallpaper.

Viscosity is due to the movement of molecules in a fluid between layers moving at different speeds. Molecules that transfer from slow-moving layers to faster-moving layers 'drag' on the faster moving layers. Thus viscosity is a form of friction between layers in a fluid moving at different speeds. Fast-moving layers in a moving fluid are dragged by slower-moving layers, which are dragged by layers moving even slower and so on. For example, when a fluid flows through a pipe, the inside surface of the pipe drags on fluid nearby, which drags on fluid further away, and so on.

A health problem

Narrow arteries restrict blood flow and cause the heart to work harder to keep the blood circulating round the body. The flow rate through a pipe of diameter D is proportional to D^4. This means that if an artery is narrowed by internal deposits to half its normal width, the flow rate would be reduced to one-sixteenth of its normal rate unless the heart works 16 times harder. In reality, the heart would be unable to work that much harder and so the flow rate would be reduced considerably.

QUESTIONS

8 (a) *Describe and explain what is observed when a soluble aspirin is dropped into a glass of water.*

(b) A drinking glass contains some lemonade and a slice of lemon. The lemon slice is then removed. Explain why the lemonade still tastes of lemon even though the slice has been removed.

9 (a) Place these liquids in order of increasing viscosity:
   cream   syrup   tar   water

   (b) Correction fluid is used to erase errors on paper. Why is it necessary to shake the bottle first if it has not been used for some time?

   (c) Describe a situation where a liquid becomes more viscous when it is warmed.

Pressure

A fluid will generally flow from high pressure to low pressure, driven by the force due to the pressure difference. For example, in most homes, cold water is supplied to a cold water tank via feed pipes connected to a main water pipe outdoors under the ground. Water is pumped into the main pipe at high pressure at a local pumping station and flows along the pipe into the homes when required.

Pressure is caused when an object exerts a force on a fluid. The fluid in a syringe squirts out of the nozzle when a force is applied to the end of the syringe. The greater the force, the faster the fluid squirts out. The narrower the syringe, the easier it is to use, because less force is needed to create a certain pressure with a narrow syringe, compared with a wide syringe. The pressure in the syringe is the force per unit area acting on the fluid. The greater the force, or the smaller the area of the object, the higher the pressure created in the fluid.

> Pressure = force per unit area

The unit of pressure is the **pascal** (Pa), which is equal to the pressure exerted by a force of 1 newton acting on an area of 1 square metre, perpendicular to the area.

HYDROSTATIC PRESSURE

The force of the Earth's gravity causes pressure in a fluid.
The pressure in a liquid in an open container increases with
depth below the surface. This is why water runs out of a sink
more slowly as the sink empties. The pressure of the water at
the plughole depends on the depth of water in the sink. As the
depth becomes less, the pressure becomes less and the rate of
flow drops.

Consider a liquid of density ρ in an upright cylinder open at the
top, as in Figure 9.4.

Figure 9.4 Hydrostatic pressure.

1 *The weight of the liquid in the cylinder = mg, where m is the
   mass of the liquid in the cylinder.*
2 *The pressure on the base of the cylinder =*

$$= \frac{\text{the weight of the liquid}}{\text{the base area } A} = \frac{mg}{A}$$

3 *The volume V of the liquid in the cylinder = Ah, where h is the height of the liquid in the cylinder.*

The pressure at the base = $\dfrac{mgh}{Ah} = \dfrac{mgh}{V} = \rho gh$

since the density of the liquid $\rho = \dfrac{m}{V}$

4 *Thus the pressure p due to a column of liquid of height h and density ρ is given by the equation*

$$p = \rho gh$$

Worked example

g = 9.8 m/s^2

Calculate the pressure due to a depth of 5.0 metres in water. The density of water = 1000 kg/m^3.

SOLUTION

p = $h\rho g$ = 5.0 × 1000 × 9.8 = 49 000 Pa.

About high blood pressure

Blood pressure is traditionally measured using an instrument called a **sphygmomanometer,** which consists of an inflatable cuff attached to a hand pump and a mercury-filled container. The cuff is wrapped round the upper arm and inflated so that it cuts off the blood supply to the lower arm. The pressure in the cuff causes the mercury to rise from the container up a vertical glass tube fixed to the container. The cuff pressure is then released and the mercury level drops. The pulse at the wrist can be detected when the cuff pressure drops sufficiently to allow blood flow to resume. The height of the mercury in the glass tube at this point is a measure of the maximum or 'systolic' blood pressure. Blood pressure is usually expressed in millimetres of mercury (abbreviated mmHg as the chemical symbol for mercury is Hg).

Figure 9.5 Measuring blood pressure.

Blood pressure varies due to the beating of the heart. A person's blood pressure is usually measured and recorded as the systolic pressure and a lower value known as the 'diastolic pressure' which is when the flow becomes quieter. The blood pressure of a typical young person is 120/80 mmHg, which means that the systolic pressure is 120 mmHg and the diastolic pressure is 80 mmHg. A systolic pressure reading considerably in excess of this indicates high blood pressure and the need for medical treatment.

To convert mmHg to pascals, use the formula $p = \rho g h$ where h is the reading converted from mm to metres, ρ is the density of mercury, which is 13 600 kg/m^3, and $g = 9.8$ m/s^2.

For a reading of 120 mmHg, $h = 0.120$ metres. Therefore:

$$p = 0.120 \times 13\ 600 \times 9.8 = 16\ 000 \text{ Pa.}$$

GAS PRESSURE

The pressure of a gas is due to gas molecules hitting the container surface and rebounding. In a gas, the average separation of the molecules is much larger than the size of a molecule. In comparison, the molecules in a liquid or in a solid are always in contact with each other. Gas molecules move about at high speeds, colliding with each other and the container at random. The motion of gas molecules in a box is like the motion of squash balls flying about in a squash court, hitting the walls and rebounding and occasionally hitting each other. The pressure due to the effect of gravity on the gas is much much smaller than the pressure due to the repeated impacts of the molecules on the container. The force due to the impacts is smoothed out because the number of molecules per second hitting the surface of the container is very large.

gas molecule

surface

Figure 9.6 Gas pressure.

The first direct evidence for the existence of molecules was obtained by the eighteenth-century Scottish botanist Robert Brown. He used a microscope to observe tiny pollen grains floating in air and he saw that the grains quivered as they moved about at random. He reasoned that their quivering haphazard motion was because each grain was bombarded unevenly by fast-moving molecules. The uneven impacts by molecules too small to see were forceful enough to push each tiny grain about at random.

The pressure of a gas in a sealed container increases if the gas is heated. This happens because the molecules of the gas move about faster when the gas is heated. Thus the impacts on the container surface are more forceful and more frequent and so the pressure of the gas is greater. If the gas is cooled, the pressure falls as the molecules move more slowly and their impacts are less forceful and less frequent. The pressure would be zero if the gas was cooled to absolute zero, which is −273°C. (See p. 67.)

The pressure of a gas at constant temperature increases if the volume of the gas container is reduced. For example, if the air in a cycle pump is trapped by blocking the outlet and then compressed by pushing the handle down the barrel, the pressure of the air in the pump increases. The reason why this happens is that the air molecules hit the container surface more often in a smaller space because they have less distance to travel between successive impacts with the container surface.

QUESTIONS

$g = 9.8 \, m/s^2$

10 In terms of pressure, explain why:
   (a) air supplied to a diver under water must be at high pressure
   (b) expanding your chest causes you to draw air into your lungs
   (c) the wall of a dam is thicker at the base than at the water surface.

11 (a) Explain why the pressure in a vehicle tyre is greater immediately after a car journey than before.
   (b) The brake system of a vehicle contains a brake fluid. Explain why it is important that the brake system contains no air bubbles.

12 (a) Calculate the hydrostatic pressure at the bottom of a swimming pool of depth 2.0 m. The density of water is 1000 kg/m³.
   (b) A blood pressure monitor attached to a patient gives a reading of 120 mmHg. Convert this reading to pascals. The density of mercury is 13 600 kg/m³.

10 THINGS TO REMEMBER

1 An **element** *is a substance that cannot be broken down into other substances. There are 92 naturally occurring elements.*

2 An **atom** *is the smallest part of an element that is characteristic of the element. A* **molecule** *is two or more atoms joined together.*

3 *Every atom has a positively charged* **nucleus** *composed of* **protons** *and* **neutrons** *(except for the hydrogen atom which has a single proton as its nucleus). The nucleus of an atom is surrounded by negatively charged* **electrons**.

4 **Isotopes** *are atoms with the same number of protons and different numbers of neutrons.*

5 *Different types of* **bond** *between atoms include the ionic bond, the covalent bond, metallic bonds and molecular bonds.*

6 **Elasticity** *is a property of any solid that regains its shape after being distorted.*

7 *Atoms in a* **crystal** *are in a regular pattern. A* **polymer** *consists of long-chain molecules. The molecules in an* **amorphous solid** *are fixed in random positions.*

8 **Diffusion** *is the spreading out of a liquid or gas into another liquid or gas.* **Viscosity** *is internal friction in a liquid in motion.*

9 **Pressure** *is force per unit area. The unit of pressure is the pascal (Pa).*

10 *The pressure at the base of a column of liquid of height h at rest = ρgh, where ρ is the density of the liquid.*

10

The age of new physics

In this chapter you will learn:
- *how Einstein revolutionized science a century ago*
- *about the theory of relativity and the quantum theory, the two great theories of modern physics*
- *why quantum effects are uncertain*
- *what is meant by time travel.*

This chapter is about the two great theories of modern physics, the quantum theory and the theory of relativity. Both theories, undiscovered at the end of the nineteenth century, revolutionized physics in the twentieth century. The quantum theory is about observations, processes and interactions involving events at a sub-microscopic scale. Information technology and communications would not have developed without the quantum theory, which provides the theoretical basis of devices such as transistors and integrated circuits. Relativity theory is about the nature of space, time, mass and energy. Nuclear power and discoveries such as quarks and black holes are consequences of the theory of relativity. In this chapter we will look at the ideas of quantum theory and the theory of relativity before meeting nuclear power, quarks, black holes and other big discoveries in the final chapters of this book.

Quantum theory

New discoveries in physics at the start of the twentieth century
created a revolution in the subject as previous assumptions about
the nature of matter and radiation were overthrown. The discovery
of photoelectricity (see p. 146) led to the conclusion that light
consists of 'wave packets' or **photons**. The smallest quantity of
light is therefore the photon, and a light beam consists of many
photons. The discovery of the electron produced the conclusion that
the smallest quantity of electric charge is the charge carried by the
electron. An electric current consists of many electrons moving on
average in the same direction. Quantities such as electric charge and
light were previously thought of as quantities that flow or spread
out continuously. However, the new physics showed that these
quantities are multiples of basic amounts, the basic amount referred
to as a quantum. Thus the quantum of charge is the charge carried
by the electron. The quantum of light is the photon. These ideas
collectively became known as the **quantum theory**. In the next few
pages, we will look at some of the key rules of quantum theory.

WAVE MECHANICS

We have seen that light has a wave-like nature because it produces
an interference pattern when it passes through two closely spaced
slits. Light also has a particle-like nature because it consists of
wave packets or photons. If light can have both a wave-like nature
and a particle-like nature, can a matter particle such as an electron
possess a wave-like nature? This question was considered by the
French physicist Louis de Broglie, who in 1923 put forward the
hypothesis that particles have a dual nature: wave-like or particle-
like, according to the situation they are in. He proposed that the
wavelength λ of the waves and the speed v of the particle were
linked through the equation

$$\lambda = \frac{h}{mv}$$

where h is the Planck constant and m is the mass of the particle. This is known as the **de Broglie wavelength** of a particle.

Within a few years de Broglie's strange ideas were shown to be true when it was discovered that a beam of electrons directed at a thin metal foil is scattered in certain directions only. The same type of effect happens when a beam of light is directed at a set of closely spaced slits; light spreads out on passing through each slit. This effect, known as **diffraction**, happens when any type of wave passes through a gap. The diffracted light waves reinforce in certain directions only. Diffraction is a wave property and reinforcement is an example of interference (see p. 137), which is also a wave property. Electrons in a beam thus behave as waves when the beam passes through a thin metal foil, since they emerge in certain directions only, just as a light beam does when aimed at a set of parallel slits.

Insight

When sea waves enter a harbour through a narrow entrance, they diffract and spread out as they cross the harbour.

Figure 10.1 Electron diffraction.

The **electron microscope** is much more powerful than a light microscope. Electrons in a beam in a vacuum tube pass through a thin specimen and are scattered by it. Magnetic 'lenses' are used to focus the scattered electrons onto a fluorescent screen where they form a magnified image of the specimen. Much more detail can be seen in the image than in a light microscope. The amount of detail possible in a microscope depends on how much diffraction occurs when the waves pass through the lenses. The smaller the wavelength of the waves or the bigger the gap they must pass through, the less the diffraction that occurs and the greater the detail that can be seen in the image. The electrons in the beam have a much shorter wavelength than light waves and so much more detail is seen.

ENERGY LEVELS AND LINE SPECTRA

Wave mechanics is used to explain why the electrons in an atom occupy 'shells' and why an electron in an atom has a fixed amount of energy depending on the shell it occupies. The energy of an electron in a shell is called an energy level. An electron trapped in an atom does not have enough kinetic energy to escape from it. The force of electrostatic attraction between the electron and the nucleus prevents the electron from leaving the atom. In effect, the electron is trapped in a 'well' (see Figure 10.2). The de Broglie wavelength of an electron in the well must be such that a whole number of wavelengths fits across the well. Because the speed of an electron and its de Broglie wavelength are related by de Broglie's equation $\lambda = h/mv$, it follows that the speed of an electron in a well can only have certain values. Thus the kinetic energy of the electron can only have certain values. As a result, the energy of the electron in the well is at certain levels only. The exact values for these levels depends on the shape and depth of the well, which depends on the charge of the nucleus and how many other electrons are present. The key point, however, is that the energy of an electron in a confined space is 'quantized', not continuous. Each energy level corresponds to a 'shell' surrounding the nucleus where an electron at that energy level is located.

The hydrogen atom is the simplest atom as it consists of a single proton as the nucleus and a single electron. The ideas above give the energy levels of the hydrogen atom at values of E_1/n^2, where

$n = 1$ for the innermost shell, $n = 2$ for the next shell, etc, and E_1 is the energy level of the innermost shell.

Figure 10.2 Energy levels.

Insight

The waves in Figure 10.2 are called 'standing waves' because they don't travel across the well and fit exactly across it. You can create standing waves on a guitar string by plucking it gently at any point such that the distance to the furthermost end is the same as, or an odd number of times, the distance to the nearer end.

The energy levels of an atom determine the **line spectrum** of light that it emits. A line spectrum consists of a pattern of well-defined bright coloured lines, each line due to light of a certain wavelength from the atom. When an atom emits light, an electron in a shell moves to a different shell at a lower energy level. The electron loses energy and releases a photon, which carries away the energy lost by the electron. Photons with the same energy all have the same wavelength, so the spectrum of light from an atom consists of well-defined wavelengths. The spectrum of light can be seen by using a prism to split the light into coloured lines, each line being due to light of a certain wavelength. The pattern of coloured lines is the line spectrum of the type of atom that emits the light. Each type of atom produces its own line spectrum because the energy levels of each type of atom are unique to that type of atom. The hydrogen spectrum consists of lines that match exactly electron transitions between energy levels given by the formula on p. 183.

Figure 10.3 A line spectrum.

THE UNCERTAINTY PRINCIPLE

So far we have seen that quantum theory means that energy, the capacity to do work, is not a continuous quantity but is quantized. Everyday experience suggests otherwise, as a light ray seems to be a steady stream of energy and the speed of any normal moving object is not restricted to allowed values only. Yet particles at the sub-atomic scale must obey the rules of the quantum world. The reason why we do not recognize the quantum world in our everyday experiences is that such experiences are due to very large numbers of particles, each with its own energy. We don't see individual photons of light because even the faintest image seen by the eye is due to huge numbers of photons.

Quantum effects are observed where the action of making an observation affects the outcome of the observation. For example, when a thermometer is used to measure the temperature of an object, energy transfer takes place from the object to the thermometer. For a large object, this transfer makes no measurable difference to the temperature of the object. However, if the object were sufficiently small, the action of making the measurement would alter its temperature. Another example is the determination of the speed of an object. This could be done by timing the object as it passes two fixed positions. The object would need to be observed as it passes each position. On a large scale, the act of observing the object would not affect its speed. However, if the object is a sub-atomic particle, photons used to observe it could change its speed when they are scattered by it. Thus the action of making an observation on this scale determines what is observed.

Insight

Imagine a radar gun is used to measure the speed of a gas molecule. The radar pulses would bounce off the molecule and change its momentum and therefore its speed would be changed.

Quantum effects are thus uncertain and need to be described as such. These ideas were put on a formal basis by the German physicist Werner Heisenberg, who proved that there is a fundamental uncertainty in the state of a particle in that its speed and its position cannot be determined simultaneously. The less uncertain the position of a particle is, the more uncertain its speed is. The less uncertain the speed of a particle is, the more uncertain its position is. Heisenberg's **uncertainty principle** leads to the conclusion that the shorter the duration of a process, the more uncertain the energy of the particles involved. Clearly, energy in the quantum world is a much more mysterious quantity than in the world of machines and engines.

QUESTIONS

1 *In each of the following experiments, state whether the electron behaves as a wave or a particle.*
   (a) An electron in a beam is deflected by a magnetic field.
   (b) An electron in a beam passes through a metal foil and is diffracted in a certain direction.
   (c) An electron in an atom is knocked out of the atom by a photon.

2 *A rainbow shows a continuous spectrum of colour.*
   (a) Describe the difference between a continuous spectrum and a line spectrum.

(b) Explain why an optical line spectrum indicates that the electrons in the light source are at well-defined energy levels.

3   The diagram below shows three of the energy levels of the electrons in a certain atom, C being the lowest.
   (a) List the three different downward electron transitions that are possible in this atom.
   (b) Which of the three electron transitions produces the smallest photon energy?

4   The speed of the electrons in an electron microscope can be increased by increasing the voltage applied to the microscope. The voltage is increased so the speed of the electrons is doubled.
   (b) How is the de Broglie wavelength of the electrons in the beam changed?
   (c) How is the final image affected by this increase of speed?

Relativity

No object can be made to travel faster than light. The speed of light in a vacuum is the ultimate speed limit. Albert Einstein was the first person to realize the fundamental significance of the speed of light. Before Einstein, scientists had attempted to find out if the speed of light was affected by the speed of the observer or the light source. Newton's laws of motion predicted that if an object is thrown forwards at a speed of 10 m/s from a vehicle moving in a straight line at 30 m/s, the speed of the object relative to the ground is 40 m/s. Physicists before Einstein thought the same rule should apply to light from a moving light source. They knew that the speed of light in a vacuum is 300 000 km/s and so they reasoned that a light beam on a moving source aimed in the forward direction of the source ought to travel faster than that. Experiments were devised to test this prediction but the speed of light was found to be the same, regardless of the speed of the light source. No one could explain this result satisfactorily, until Einstein provided the answer in 1905 when he published the theory of relativity.

At that time Einstein was no more than a junior scientist at the Swiss Patent Office. His official duties were not too demanding and he was able to develop his ideas in his spare time. His theory of relativity started with just two assumptions:

1 *The speed of light is the same for any observer, regardless of the speed of the source or the speed of the observer.*
2 *All motion is relative.*

The first assumption cuts out the argument about whether the speed of light depends on the speed of the light source. The second assumption means that all physical laws expressed in mathematical form must be the same for all observers moving at constant velocity relative to each other. Einstein put these assumptions forward as his starting points and then worked mathematically through their consequences. We shall leave aside the mathematical details here and look at why the predictions made by Einstein revolutionized our understanding of space and time. Two of the predictions were:

▶ *The observed length of a moving rod is shorter, the faster the rod travels towards or away from the observer.*
▶ *A moving clock runs more slowly than a stationary clock.*

Both these effects require objects to be moving at speeds approaching the speed of light. Experiments using particle beams at speeds approaching the speed of light have confirmed Einstein's predictions. For example, the lifetime of fast-moving unstable particles called muons is longer than if they were stationary. Another example is a twin who travels at high speed away from the Earth and back again. On return, he or she would find that he or she is younger than the stay-at-home twin, because a traveller's clock runs more slowly than a stationary clock.

Insight

Fast-moving muons are created when particles from space (cosmic rays) collide with atoms in the upper atmosphere. These muons last much longer than slow muons and reach the ground in significant numbers.

Two further predictions made by Einstein in his 1905 theory of relativity revolutionized our understanding of mass and energy.

▶ *The mass of a moving object increases with its speed. If an object is accelerated to a greater speed, its mass increases. Einstein showed that the mass would be infinite at the speed of light; hence the conclusion that no material object can be made to travel as fast as light.*
▶ *Energy E can be converted into mass m and vice versa in accordance with Einstein's famous equation*

$$E = mc^2$$

where c is the speed of light in a vacuum. For example, the conversion of 4 tonnes (= 4000 kg) of matter into energy would release about 4×10^{20} J of energy, enough to meet the annual demand for energy for the entire world. Check the calculation yourself, given m = 4000 kg and c = 3×10^8 m/s.

Albert Einstein 1879–1955

Einstein did not distinguish himself academically at secondary school in Munich. His family had moved to Munich from Ulm a year after his birth and moved to Milan 14 years later. Einstein disliked the regimented school system that prevailed in Germany at that time and he left school prematurely. His family recognized his unusual talents in mathematics and physics and arranged for him to complete his secondary education in Zurich. In 1896 he gained a place to study at the Swiss Federal Institute of Technology (known as the E.T.H.) in Zurich and graduated with a physics degree four years later. He developed a reputation for asking awkward questions which his professors could not always answer. Two years after leaving E.T.H., he secured a post in the Swiss Patent Office at Berne. In his spare time he worked at his ideas on physics, with the support of his wife and his friends. In 1905 he published papers on the photon theory of radiation and the theory of relativity, now referred to as **special relativity**. He was appointed to a part-time post at the University of Berne in 1907 and became a professor in 1909. In 1914, after a brief spell in Prague, Einstein took up the post of Director of the Kaiser Wilhelm Institute for Physics in Berlin. He published the **general theory of relativity** in 1916, showing that gravity is a feature of space and time. He became a household name when his prediction that gravity bends space was confirmed in 1919 by the British astronomer Sir Arthur Eddington. In 1933 Einstein was forced to emigrate from Germany and he spent the rest of his life in the USA.

Experimental evidence in support of the theory of special relativity was found in the discovery that the specific charge of the electron (i.e. its charge/mass value) was smaller, the greater the speed of the electron. Further tests showed that the mass of an electron did indeed increase with speed, as Einstein had predicted. The masses of atoms were measured accurately and it was discovered that mass was lost when a radioactive change took place and that the energy released matched the mass loss in accordance with $E = mc^2$. More dramatically, nuclei can be made to release energy and this process can be caused on a scale that could wipe out the human race if ever nuclear weapons were to be used in a war. In addition, physicists have discovered that high-energy photons are capable of turning into matter in the form of particles and antiparticles. The reverse process is also possible where a particle and an antiparticle annihilate each other and turn into photons. All these processes require the use of $E = mc^2$. Perhaps the most amazing aspect of Einstein's work is that after a century of using $E = mc^2$, the mechanism behind the conversion of mass and energy is still not clearly understood. With these thoughts in mind, we will move on in the next chapter to look at what is now known about matter and its properties at an even smaller scale than the nucleus.

QUESTIONS

$c = 3 \times 10^8 \, m/s$

5 (a) An atom of mass 2×10^{-28} kg releases a photon of light of energy 1×10^{-19} J. Show that the mass loss of the atom is negligible.

   (b) A nucleus in an atom of mass 2×10^{-28} kg releases a photon of energy 1×10^{-13} J. Show that the mass of the nucleus decreases by about 0.5% as a result of this process.

6 The Sun radiates energy at a rate of 4×10^{26} watts. Calculate the amount of mass converted into energy by the Sun every second.

10 THINGS TO REMEMBER

1 *Light and all other forms of electromagnetic radiation consist of* **photons,** *which are packets of electromagnetic waves.*

2 *Matter particles have a wave-like nature as well as a particle-like nature.*

3 *The greater the momentum of a particle, the smaller its de Broglie wavelength.*

4 *The electrons in an atom occupy shells at fixed energy levels.*

5 *An atom emits a photon of light when one of its electrons moves to a shell of lower energy.*

6 *The* **uncertainty principle** *tells us that the momentum and the position of a particle cannot be determined simultaneously.*

7 *The* **speed of light** *is independent of the speed of the light source and the speed of the observer.*

8 *A moving clock runs more slowly than a stationary clock.*

9 *No object can travel faster than c, the speed of light in free space.*

10 *Energy E and mass m are interchangeable on a scale given by* $E = mc^2$.

11

The structure of matter

In this chapter you will learn:
- *what radioactivity is and what causes it*
- *what is inside the nucleus of the atom*
- *about quarks and leptons, the fundamental particles from which all matter is composed.*

The search to discover the fundamental structure of matter has been the central theme of science for the past few centuries and is a quest that continues today. New discoveries build on previous discoveries and existing theories, enabling existing theories to be refined or new ideas developed. The ideas about atoms that were first put forward thousands of years ago by Democritus in Ancient Greece were rediscovered by John Dalton in the early nineteenth century. Dalton's theory that atoms are indivisible and indestructible worked well for almost a century until the discovery that atoms contained much lighter negative particles, which became known as electrons. Experiments by Ernest Rutherford in the first decades of the twentieth century showed that every atom contains a positive nucleus where most of its mass is located. Further experiments led to the conclusion that the nucleus is composed of protons and neutrons, and that the electrons move round the nucleus. In this chapter we will look at the discoveries that led to this picture of the atom, before moving on to look at the astonishing picture revealed by physicists probing matter on incredibly small scales. We know now that protons and neutrons are composed of particles called quarks.

We know that for every type of particle there is a corresponding type of antiparticle. Like opening Pandora's box, which has a box inside a box inside a box and so on, the search to discover the structure of matter has uncovered structure at successively smaller scales. Now the search is on to find out if quarks and electrons are manifestations of nature on an even smaller scale.

Inside the atom

New discoveries often lead to new inventions, which then lead to more new discoveries. For example, in the mid-nineteenth century, Michael Faraday's discovery of electromagnetic induction led to the invention of a range of electromagnetic devices such as the dynamo (see p. 116), the transformer (see p. 118) and the induction coil. The induction coil is like a transformer, except that the primary coil is connected to a battery in series with a 'make-and-break' switch. In operation, the 'make-and-break' switch repeatedly switches the current through the primary coil on and off. Each time the primary current is switched on or off, the changing magnetism through the secondary coil causes an induced voltage across the secondary coil. The secondary coil has many more windings on it than the primary coil. Consequently, a very large voltage is induced in the secondary coil each time the make-and-break switch opens or closes.

Sir William Crookes used the induction coil in the mid-nineteenth century to investigate if gases conduct electricity at very low pressures and high voltage. He found that gases do conduct electricity when subjected to high voltage at very low pressures. More importantly, he found that light is emitted by the gas in these conditions. Neon tubes and other colourful illuminated displays are the result of Crookes' discovery that the colour of light emitted depends on the type of gas in the tube. Crookes carried out further research into the nature of the particles responsible for carrying electricity in gases. It was already known that conduction in a liquid solution, such as salt water, is due to the presence of positive and negative ions in the solution. When two electrodes connected to a battery are placed in the solution, the positive ions are attracted to the negative electrode and the negative ions are attracted to the positive electrode. Crookes worked out that a gas at low pressure conducts for the same reason, namely the presence of positive and negative ions in the gas. However, unlike in a liquid solution, ions are created in a gas only when it is at low pressure and a high voltage exists across it. The high voltage across the gas causes electrons to be torn out of individual gas atoms, leaving the atoms as positive ions to be attracted towards the negative electrode in the gas while the electrons are attracted towards the positive electrode. At the time, Crookes realized that positive and negative ions were created in the gas and he reasoned that light was emitted by the gas when a positive ion and a negative ion collide and recombine to form an atom again. The negative ions were later identified as electrons.

Insight

An ion is a charged atom. A positive ion is formed by removing an electron from an uncharged atom. A negative ion is formed by adding an electron to an uncharged atom.

THE DISCOVERY OF THE ELECTRON

Research by Joseph Thomson into conduction in gases at very low pressure led to the discovery of the electron in 1897. Thomson found that the negative ions were always the same, regardless of which gas was present. He discovered that the positive ions were gas atoms with electrons removed. Thomson made these discoveries using apparatus he designed and constructed to produce beams of ions. Thomson's apparatus was in effect the very first TV-like tube, as the tube widened out into a screen with a flat end. The screen was coated internally with a chemical that produced a spot of light where the ion beam hit the screen.

By applying an electric field or a magnetic field to the beam before it hit the screen, the ions could be deflected, and the deflection could be measured from the shift of position of the spot of light on the screen.

1   *The electric field was created between two metal 'deflecting' plates, one above the other. One plate was connected via a switch to the positive terminal of a high-voltage supply unit and the other plate to the negative terminal of the supply. The deflecting plates were positioned so that the beam passed between them. When the switch was closed, the spot on the screen was seen to deflect because the ions were attracted to the oppositely charged deflecting plate.*

2   *The magnetic field was created by passing a current through two coils of insulated wire placed one on either side of the narrow part of the tube. The deflection could be reversed by reversing the current and it could be increased by increasing the current. This principle is used in all modern TV tubes and VDUs to make an electron beam scan the screen and form a picture on the screen.*

Figure 11.1 Deflection of electrons.

Thomson found that the deflection of the negative ions did not depend on the type of gas used, whereas the deflection of the positive ions did. He realized that the negative particles are in every type of atom and referred to them as 'corpuscles of electricity' before they became known as electrons. He measured the deflection of the electrons in a beam by deflecting them with electric and magnetic fields of known strengths. From his measurements he worked out the charge/mass value of the electron (known technically as its **specific charge**). This was a significant measurement, because the specific charge of most ions had been measured decades earlier in electrolysis experiments. For example, it was known that hydrogen gas is released when electricity is passed through acidified water and that 96 000 coulombs of electric charge were needed to produce 1 gram of hydrogen gas. The specific charge of the hydrogen ion was therefore known to be 96 000 coulombs per gram, larger than the specific charge of any other ion – until Thomson discovered that the specific charge of an electron is about 1850 times greater. Thomson could not be certain if this very large value was because the electron is much lighter than the hydrogen ion or because the electron carries much more charge than a hydrogen ion. However, assuming that the hydrogen ion and the electron carry equal and opposite amounts of charge, Thomson and most of his contemporaries reckoned that the electron must be much lighter than the hydrogen atom.

Northern Lights

The Aurora Borealis, also known as the Northern Lights, is visible at night in summer in the northern hemisphere. The effect occurs because fast-moving protons and other charged particles streaming from the Sun are channelled by the Earth's magnetic field into a region in the upper atmosphere where they collide with gas atoms, which then emit light. The same effect happens in summer in the southern hemisphere.

The charge of the electron e was measured by the American physicist Robert Millikan in 1915. His value of e was very close to the now-accepted value of 1.6×10^{-19} coulombs. The mass of the electron could then be calculated from its specific charge $e/m = 1.8 \times 10^8$ coulombs per gram (= $1850 \times 96\,000$ coulombs per gram) and its actual charge $e = 1.6 \times 10^{-19}$ C. Prove for yourself that these figures give a value of 9×10^{-28} grams (= 9×10^{-31} kilograms) for the mass of the electron.

Within 20 years of Thomson's discovery, scientists had worked out that the electrons in an atom orbit a nucleus of positive charge in the atom. The discovery of the nucleus is part of the next thread in the development of our understanding of the structure of the atom.

QUESTIONS

1   *Use the words in the following list to complete the sentences below.*

    atoms   ions   light   pressure   voltage

    A gas conducts electricity if it is at low _____ and a high _____ is applied across it. Positive _____ and negative _____ are created from gas _____ . The _____ move to the oppositely charged electrode in the tube. Light is emitted if a positive _____ and a negative _____ collide and recombine.

2   *A beam of electrons hits a screen and produces a spot of light. The beam is then deflected using a magnetic field, which causes the spot to move a centimetre downwards on the screen. How would the spot be affected if the magnetic field was (a) increased in strength, (b) reversed in direction?*

3   *The specific charge of the electron is 1.76×10^{11} coulombs per kilogram. The charge of the electron is 1.60×10^{-19} C. Use this data to show that the mass of the electron is 9.1×10^{-31} kg.*

Radioactivity

Radioactivity was discovered accidentally by Henri Becquerel when, in Paris in 1896, he was conducting research into the effects of X-rays on uranium compounds. He had discovered that certain substances exposed to X-rays glow and continued to glow when the X-ray machine was switched off. He wanted to know if the reverse effect was possible, namely that a substance would emit X-rays after it had been exposed to strong sunlight. In readiness for a sunny day, he placed a wrapped photographic plate in a drawer with a small quantity of the uranium compound on it. After several dull days, he decided to develop the plate, expecting to observe no more than a faint image of the compound. He was therefore very surprised when he found a very strong image on the plate. He realized that the uranium compound was emitting some form of radiation without having been exposed to sunlight. Further tests showed that the substance emits this radiation continuously even when stored in darkness for long periods and that the radiation passes through glass, but not through metal. The substance was described as 'radioactive' because it did not need to be supplied with energy to make it emit radiation and was therefore emitting radiation actively. Becquerel continued his research on X-rays and passed the investigation of radioactivity on to his research student, Marie Curie. Within two years, Marie Curie and her husband Pierre had discovered other substances that are radioactive, including two new elements, radium and polonium. Becquerel and the Curies were awarded the 1903 Nobel prize in physics for their discoveries.

The nature of radioactivity was established by Ernest Rutherford, who showed that the radiation is produced when unstable atoms disintegrate. He used the radiation to probe the atom and deduced that every atom contains a positively charged nucleus where most of its mass is located. Further research showed that the nucleus itself is composed of neutrons and protons. In recent decades, scientists have discovered that protons and neutrons are composed of smaller particles, which are referred to as quarks. Many questions about the sub-atomic world remain unanswered and more questions arise as more discoveries are made. We will look at some of these questions later in this chapter, after we have studied radioactivity in more detail.

Radioactivity is due to the instability of a nucleus which has too many protons or neutrons. Such a nucleus becomes stable, or less unstable, by emitting one of three types of radiation, as below. All of these are capable of creating ions in air by knocking electrons out of air molecules.

1 **Alpha radiation** (α) *consists of particles, each composed of two protons and two neutrons. An α particle is emitted by a very large unstable nucleus. Alpha radiation:*
   ▷ *is easily stopped by cardboard or thin metal*
   ▷ *has a range in air of no more than a few centimetres*

> ionizes air molecules much more strongly than the other two types of radioactive radiation.

2 **Beta radiation** (β) *consists of electrons, each emitted when a nucleus with too many neutrons disintegrates. A neutron in such a nucleus suddenly and unexpectedly changes to a proton; in the process, an electron is created and instantly emitted from the nucleus. Beta radiation:*
  > *is stopped by 5–10 mm of metal*
  > *has a range in air of about 1 metre*
  > *ionizes air molecules less strongly than α radiation.*

3 **Gamma radiation** (γ) *consists of high-energy photons. A photon is a packet of electromagnetic waves. A gamma photon is emitted from a nucleus with surplus energy after it has emitted an α or a β particle. Gamma radiation:*
  > *is stopped only by several centimetres of lead*
  > *has an infinite range in air*
  > *ionizes air molecules very weakly.*

Figure 11.2 Radioactivity.

Insight

Most nuclei are stable. If they were not, they would have disintegrated long ago.

The **Geiger counter** is used to detect radioactive radiation. This device was invented by Hans Geiger, who worked with Rutherford. It consists of a tube connected to an electronic counter. Each time an ionizing particle enters the tube, the counter registers it as a single count and a click is heard. The tube itself is sealed and hollow, with a thin window over one end and a metal rod along its axis. The tube contains gas at very low pressure. With several hundred volts between the rod and the tube, an ionizing particle entering the tube ionizes the gas atoms which then ionize more atoms, etc. The gas becomes conducting for a fraction of a second, causing a tiny pulse of electricity to pass through the counter and be registered.

Figure 11.3 Using a Geiger counter.

To use a Geiger counter:

1 *The counter is set to zero.*
2 *The 'start count' switch is pressed and a stopwatch is started at the same time.*
3 *The 'stop count' switch is pressed after a measured time (e.g. 100 seconds).*

The count rate is the number of counts per second. This is calculated by dividing the number of counts by the counting time. In practice, the measurement is repeated several times to obtain an average value for the count rate.

Background radioactivity must be taken into account when measuring the count rate due to a radioactive source. Background radioactivity is caused by radioactive substances present in rocks, by the effect of cosmic radiation on the atmosphere and by radioactive pollutants released into the atmosphere, such as happened at Chernobyl in 1986. To take account of background radioactivity, the count rate is measured with the source present and without the source present. The count rate due to the source is the difference between the two measurements.

HALF-LIFE

The **half-life** of a radioactive isotope is the time taken for half the number of atoms of the isotope to disintegrate. Suppose 10 000 atoms of a certain radioactive isotope X are present initially. The number of atoms decreases

▶ *from 10 000 to 5000 after one half-life, then*
▶ *from 5000 to 2500 after a further half life, then*
▶ *from 2500 to 1250 after a further half life, etc.*

The amount of the radioactive isotope therefore decreases with time as shown in Figure 11.4, which is a **half-life curve**. Half-life values range from a fraction of a second to billions of years. For example, the half-life of uranium 238 ($^{238}_{92}$U) is about 4.5 billion years.

Figure 11.4 A half-life curve.

Radioactive disintegration is a random process. For a large number of atoms of a given radioactive isotope, the proportion that disintegrate per second is constant. This follows because of the random nature of radioactive disintegration. To appreciate this, suppose a thousand dice are rolled and all those that show a '1' are then removed. With such a large number to start with, the number of dice removed would be about 167 (= 1000/6) because, on average, one-sixth would show a '1'. If the process was repeated with the 833 remaining dice (= 1000 − 167), the number removed in this second throw would be about 139 (= 833/6). If the process were to be repeated a number of times, each time using the remaining dice, the number remaining would decrease as below.

Number of throws	0	1	2	3	4	5
Number of dice remaining	1000	833	694	578	482	402
Number of dice removed	167	139	116	96	80	67

A graph of the number of dice remaining against the number of throws is the same shape as the half-life curve in Figure 11.4. The number of dice remaining drops to about half in less than four throws. The 'half-life' of this process is about four throws therefore. You can prove for yourself that the number would drop to about 250 in about four more throws.

Insight

The Earth is about 5000 million years old. The half-life of uranium 238 is 4500 million years. If it were much shorter, all the uranium 238 in the Earth would have disintegrated long ago.

RADIOACTIVITY AT WORK

1 **Radioactive tracers** *are used in medicine and environmental technology. The tracer needs to be a radioactive isotope that emits β or γ radiation, because this type of radiation can pass*

through materials. For example, suppose a gas leak occurs in an underground gas pipe. The leak can be pinpointed by injecting a small quantity of a radioactive gas into the pipe and then moving a Geiger counter at ground level along the pipe. The source of the leak would be where the counter reads more than the background count rate. Another example, from medical physics, is in the diagnosis of a blocked kidney. The patient is given a drink of water containing a very small quantity of a radioactive tracer. A Geiger counter is held against each kidney. For a normal kidney, the count rate would rise and then fall, as the water passed through. In a blocked kidney, the count rate would rise and not fall back, as the water would not pass through the kidney. The tracer needs to have a half-life long enough for it to flow into or through the system, but short enough for its activity to become negligible soon after the measurements have been made.

2 Radioactivity is used in monitoring industrial processes, such as thickness monitoring of metal plating produced in a rolling mill. Molten metal is forced between two steel rollers and the metal cools to form a metal plate as it passes through. With a β source and a Geiger tube either side of the plate, the count rate varies according to the thickness of the plate. An increase of thickness would cause the count rate to drop and a decrease of thickness would cause the count rate to rise. The count rate signal is used to control the pressure on the steel rollers. The half-life of the source needs to be long enough so that the source does not need to be replaced too often.

3 **Radioactive dating** enables the age of ancient objects to be measured. For example, living wood contains a small proportion of a radioactive isotope of carbon, $^{14}_{6}C$. When a tree dies, the proportion of this isotope gradually decreases as these atoms disintegrate with a half-life of about 5600 years. To measure the age of an ancient wooden object, the count rate due to a sample of the object is measured and compared with the count rate of an equal mass of wood from a living tree. This comparison can then be used to work out how many half-lives have elapsed since the death of the tree used to make the ancient object. If the count rate was reduced to one-quarter, the object would be 2 × 5600 years old (= 2 half-lives).

RADIOACTIVE HAZARDS

Radioactivity is dangerous to human health because ionizing radiation kills living cells and causes tumours. For this reason, the use of radioactive sources is subject to strict regulations. For example, radioactive sources must be kept in a lead container to prevent radiation from escaping. Also, radioactive sources should only be moved using a suitable handling device.

Because radioactive sources with long half-lives remain radioactive for many years, disposal of radioactive sources and products is subject to legal regulations. Radioactive waste from nuclear reactors must be stored in sealed containers at approved sites.

QUESTIONS

4 The following measurements were made using a Geiger counter and a radioactive source with a long half-life.

   ▷ Number of counts in 600 seconds with no source present = 245.
   ▷ Number of counts in 100 seconds with the source 20 mm from the Geiger tube = 455, 461, 438.
   ▷ Number of counts in 100 seconds with a 1 mm metal plate between the source and the tube = 325, 340, 315.

   (a) Calculate the background count rate in counts per second.
   (b) Show that the corrected count rate due to the source without the plate present has an average value of 4.10 counts per second.
   (c) Show that the corrected count rate due to the source with the plate present has an average value of 2.86 counts per second.
   (d) Show that 70% of the radiation incident on the plate passes through it.
   (e) What type of radiation is emitted by the source?

5 A certain radioactive isotope has a half-life of 8 hours. A solution containing 500 million atoms of this isotope is prepared. How many atoms of this isotope have not disintegrated after
   (a) 8 hours, (b) 24 hours, (c) one week?

Quarks and leptons

Antimatter was predicted by the British physicist Paul Dirac in 1926. He reckoned that for every type of particle, there is a corresponding antiparticle with identical mass and the opposite type of charge, if the particle is charged. He predicted that a particle and its antiparticle can be created as a pair and they can annihilate each other. Within a few years, the American physicist Carl Anderson had detected the positron, the antiparticle of the electron. Anderson first observed the track of a positron in a device known as a cloud chamber. A charged particle passing through a cloud chamber leaves a track because it creates ions along its path. The ions cause tiny droplets to form from the vapour in the chamber, leaving a visible track. Anderson recognized that the track was produced by a particle like the electron, but the magnetic field applied to the chamber made the track curve in the opposite direction to that expected for an electron.

Insight

An antihydrogen atom has a nucleus consisting of an antiproton with a positron moving around it. If a rocket engine could be made to work by annihilating hydrogen and antihydrogen, it would release over a hundred million times as much energy per kilogram of fuel as an ordinary rocket engine.

Cosmic radiation consists of particles that enter the Earth's atmosphere from space. These particles are produced by the Sun or other stars. They crash into the nuclei of atoms in the atmosphere, creating cascades of particles at high speed that can reach the Earth's surface. As well as positrons, other particles called pi-mesons (or pions) were discovered. These particles were called mesons because they were found to be 'middle-weight' particles, heavier than electrons but lighter than protons. Further investigations revealed the existence of 'strange' particles that are created in particle pairs, rather than particle–antiparticle pairs, and which decay into pions and protons.

Accelerators have been constructed to create and study these new types of particle and antiparticle. In essence, an accelerator is an evacuated tube containing electrodes which are used to accelerate charged particles such as electrons or protons to speeds approaching the speed of light. These charged particles are then directed in a narrow beam at a target. Some of the charged particles collide with the target nuclei to create mesons and other short-lived particles and antiparticles from the energy of the charged particles in the beam. Using these accelerators, physicists have discovered a large number of short-lived particles and antiparticles, charged and uncharged, with a range of masses. These newly discovered particles and antiparticles were found to fit patterns that could be explained by assuming that protons and neutrons are composed of three elementary particles which became known as **quarks**.

The quark model is based on the following assumptions:

1 *There are six different types of quark: the up quark, the down quark, the strange quark, the charmed quark, the bottom quark and the top quark (referred to as u, d, s, c, b and t for brevity).*

 charge $= +\frac{2}{3}$e

 charge $= -\frac{1}{3}$e

up

down

Introducing the quarks

charge $= +1$

charge $= 0$

inside a proton

inside a neutron

Figure 11.5 Quarks.

2 *For every type of quark, there is a corresponding antiquark.*
3 *Quarks combine in threes to form particles like the proton and the neutron. Antiquarks also combine in threes to form antiparticles like the antiproton and the antineutron. Such composite particles are collectively referred to as* **baryons**.
4 *A* **meson** *consists of a quark and an antiquark.*

In terms of the charge of the electron, the u, c and t quarks each carry a charge of $+\frac{2}{3}e$ and the other three quarks carry a charge of $-\frac{1}{3}e$. An antiquark carries an equal and opposite charge to its corresponding quark. The symbol for an antiquark is the same as for a quark but with a bar over the top. For example, $\overline{d}$ (pronounced 'dee bar') represents the symbol for a down antiquark. So:

▶ *A proton is composed of two up quarks and a down quark.*
▶ *A neutron consists of an up quark and two down quarks.*
▶ *A pion consists of an up or down quark and an up or down antiquark.*
▶ *Strange particles contain strange quarks or antiquarks.*

The creation of pions and strange particles and antiparticles can be explained using the quark model. For example, if a proton at high speed collides with another proton, the following interaction could take place:

proton + proton → positive pion + proton + neutron

In quark terms:

uud + uud → u$\overline{d}$ + uud + udd

In the collision, a down quark and a down antiquark are created from the kinetic energy of the high-speed proton. The quarks and the antiquark regroup to form a positive pion, a proton and a neutron.

The quark model was confirmed by physicists using the Stanford Linear Accelerator to accelerate electrons to speeds within a tiny fraction of the speed of light and use them to bombard a target. These electrons were scattered by the target nuclei in directions corresponding to three hard centres in every neutron and proton.

Where do electrons fit into this model? The answer is that they do not. Electrons, positrons and certain other particles and antiparticles are thought to be elementary in the sense that they are not composed of smaller particles. These particles and antiparticles are collectively referred to as **leptons**. All particles not in the quark family belong to the lepton family. Quarks and antiquarks might themselves be composed of even smaller particles. The difference of charge between the up quark and the down quark is equal to the charge of the electron. This perhaps suggests that there is a deeper link between quarks and leptons. The Large Hadron Collider at CERN, the European Centre for Nuclear Research, might provide some answers in due course. This accelerator is designed to collide protons at energies more than 20 times greater than the biggest accelerators in current use. The mechanism linking mass and energy might at last be uncovered, a century after Einstein discovered the link.

Insight

Quarks are the building blocks of protons and neutrons. Electrons and all other leptons are thought to be elementary.

QUESTIONS

6   What is the total charge of a baryon that consists of (a) an up quark, a down quark and a strange quark, (b) three up quarks, (c) three down quarks?

7   What is the total charge of a meson that consists of (a) an up antiquark and a down quark, (b) a strange antiquark and a down quark, (c) a down antiquark and an up quark?

8   Write down the quark composition of (a) a proton, (b) a neutron.

9   When a beta particle is emitted from an unstable nucleus, a neutron in the nucleus changes into a proton. What change in the quark composition of the neutron takes place in this process?

10 A proton moving at high speed collides with a neutron to create a strange quark and a strange antiquark. The quarks and the

antiquark regroup to form three composite particles which include a strange meson, a strange baryon and a third particle. Identify the quark composition of the third particle.

proton + neutron → strange meson + strange baryon + ?
uud   udd   d$\bar{s}$ sdd

10 THINGS TO REMEMBER

1 A discharge tube such as a neon tube glows with light when a current passes through it.

2 A beam of electrons can be deflected by an electric or a magnetic field.

3 The atoms of a radioactive substance each have an unstable nucleus which can spontaneously emit radiation to become stable or less unstable.

4 The three types of radiation from radioactive substances, in order of increasing penetrating power, are alpha, beta and gamma radiation.

5 The radiation from a radioactive substance is harmful because it damages or kills living cells due to its ionising effect.

6 The half-life of a radioactive isotope is the time it takes for half its atoms to disintegrate.

7 Radioactive substances have many uses, including in tracing the flow of a substance and the dating of ancient objects.

8 For every type of particle there is a corresponding anti particle that has the same mass and an equal and opposite charge. When a particle and an antiparticle collide, they annihilate each other and produce radiation.

9 There are six types of quarks. The up, charmed and top quark each carry a charge of $+\frac{2}{3}$ e. The down, strange and bottom quark each carry a charge of $-\frac{1}{3}$ e.

10 A proton consists of two up quarks and a down quark. A neutron consists of one up quark and two down quarks.

12

....

Energy from the nucleus

In this chapter you will learn:
- *about nuclear fusion and nuclear fission*
- *how a nuclear reactor works*
- *about the fast breeder reactor*
- *about the benefits and drawbacks of nuclear power.*

No other branch of science has so far had such an impact on the human race as nuclear physics. Maybe this is not surprising, as nuclear weapons have the capacity to destroy the human race. Yet before the discovery of nuclear fission, Lord Rutherford reckoned that obtaining energy from the nucleus on a large scale was 'moonshine'. Nuclear fission, the splitting of a large nucleus, was discovered by Otto Hahn and Fritz Strassmann in Berlin in 1938. Physicists around the world recognized the significance of the discovery as the energy released in this process is far greater than in any other. The British and US governments saw that if Germany were to become the first nation to possess nuclear weapons, this would give dreadful military power to the Nazi dictatorship there. The Manhattan Project was set up in the USA to make a nuclear weapon, the atom bomb, before such a weapon could be made in Germany. In fact, Germany had been defeated when the first atom bombs were dropped on the Japanese cities of Nagasaki and Hiroshima, bringing the Second World War to an end. Nuclear reactors for electricity generation were subsequently developed and now account for about 8% of all electricity generation worldwide. Concerns about radioactive waste and the Chernobyl disaster caused nuclear power to fall out of favour, but the current generation of nuclear power stations will need to be replaced if fossil fuel usage is cut back to combat global warming and electricity cuts and fuel rationing are to be avoided. In this chapter we will look at the principles of nuclear fission, how nuclear reactors work and why they are likely to continue to make a significant contribution to our electricity supplies.

Nuclear fission

A radioactive substance emits radiation because the nuclei of its atoms are unstable and disintegrate at random, emitting α radiation, β radiation or γ radiation. Such changes cause energy release on a scale about a million times greater than when atoms react chemically. However, radioactive isotopes with sufficiently long half-lives are not active enough to release energy at a fast enough rate to produce electricity on a large scale. This is why Rutherford reckoned that nuclear power was unrealistic – until nuclear fission was discovered.

Nuclear fission is the splitting of a large nucleus into two approximately equal fragments. Energy is released in this process on a scale even greater than when radioactive disintegration occurs. Hahn and Strassmann proved that this process happened to the uranium isotope $^{235}_{92}U$. Further investigations showed that this isotope could be made to fission by bombarding it with neutrons, and that two or three neutrons are released when such a nucleus splits. These released neutrons are referred to as fission neutrons. Thus a chain reaction is possible in which a neutron splits a uranium 235 nucleus and two or three fission neutrons are released, which go on to split other uranium 235 nuclei, leading to more neutrons being released which then go on to split more uranium 235 nuclei. Each fission event releases energy and so an enormous amount of energy is released if the chain reaction is maintained.

incoming neutron

nucleus

neutrons released

Figure 12.1 Fission.

Insight

Dominoes standing end-on in a line fall over in sequence when the front domino is pushed. One by one all the other dominoes fall. This is like a chain reaction in which one event triggers more events which trigger more events and so on.

BINDING ENERGY

Why should a large unstable nucleus release energy when it fissions or undergoes a radioactive change? Such a nucleus wobbles about and its shape changes as it wobbles. If a neutron from outside hits it when it is distorted into two lobes still joined together, the lobes gain enough energy to separate even more. At sufficiently large separation, the electrostatic repulsion between the two lobes forces them to fly apart. The protons and the neutrons in a nucleus are normally held together by a strong attractive force called the **strong nuclear force,** which prevents the protons pushing away from one another. Fission occurs because the strong nuclear force is a short-range force, whereas the electrostatic force of repulsion is a long-range force.

To completely separate the protons and neutrons in a nucleus, work needs to be done on them to overcome the strong nuclear force. The work needed to separate a nucleus into separate neutrons and protons is referred to as the **binding energy** of the nucleus; the greater the binding energy of a nucleus, the greater the work that would be needed to separate the neutrons and the protons in the nucleus from each other.

The mass of a nucleus is less than the mass of the same number of separate neutrons and protons. For example, the mass of a helium nucleus, which consists of two protons and two neutrons, is 0.8% less than the mass of the two protons and two neutrons separated from each other. This difference is called the **mass defect** (Δm) of the nucleus and is due to the protons and neutrons binding together when the nucleus was formed. The binding energy of the nucleus can be calculated from the mass defect using Einstein's famous equation $E = mc^2$.

> Binding energy = mass defect × c^2

Nuclear masses are usually expressed in atomic mass units (u), using the carbon 12 scale (see p. 158), and the energy released by a nucleus is usually expressed in millions of **electron volts** (MeV) where 1 MeV = the energy gained by an electron accelerated through a million volts = 1.6×10^{-13} J. Using the above equation, a mass defect of exactly 1 u equates to 931 MeV of binding energy.

The **binding energy per nucleon** of a nucleus is the binding energy of a nucleus divided by the number of nucleons (i.e. protons and neutrons) in the nucleus. This quantity is a measure of the stability of a nucleus. It can be easily calculated for any nucleus $_Z^A X$ of known mass M by following the steps below.

1 *The mass defect (in atomic mass units) of the nucleus = $Zm_p + (A - Z) m_n - M$ where m_p is the mass of a proton and m_n is the mass of a neutron.*
2 *The binding energy E_b (in MeV) = $931 \times \Delta m$.*
3 *The binding energy per nucleon = E_b / A.*

Note: Z = the number of protons in the nucleus, A = the number of neutrons and protons, so $A - Z$ is the number of neutrons in the nucleus.

Worked example
The mass of a $^{235}_{92}\text{U}$ nucleus is 234.99333 u. The mass of a proton = 1.00728 u and the mass of a neutron = 1.00867 u. Calculate the binding energy per nucleon of a $^{235}_{92}\text{U}$ nucleus.

SOLUTION
$Z = 92$, $A = 235$ so the number of neutrons = $A - Z = 143$

Therefore, the mass defect = $(92 \times 1.00728) + (143 \times 1.00867) - 234.99333 = 1.91624$ u

Binding energy = $1.91624 \times 931 = 1784$ MeV

Binding energy per nucleon = $1784/235 = 7.6$ MeV per nucleon.

A graph of binding energy per nucleon against nucleon number A is shown in Figure 12.2. Remember, the greater the binding energy per nucleon of a nucleus, the more stable the nucleus is.
The graph shows that:

1 *The binding energy per nucleon increases as A increases to a maximum of about 9 MeV per nucleon at about A = 50 to 60, then decreases gradually.*
2 *The most stable nuclei are at about A = 50 to 60, since this is where the binding energy per nucleon is greatest.*
3 *The binding energy per nucleon is increased when nuclear fission of a uranium 235 nucleus occurs.*
4 *The binding energy per nucleon is increased when light nuclei are fused together.*

Insight
The greater the binding energy of a nucleon is, the more stable the nucleus is and the harder it is to split it.

When a uranium 235 nucleus undergoes fission, the two fragment nuclei each comprise about half the number of nucleons. Therefore the binding energy per nucleon increases from about 7.5 MeV per nucleon for uranium 235 to about 8.5 MeV per nucleon for the fragments. Thus the binding energy per nucleon increases by about 1 MeV for every nucleon, which means that the energy released from the fission of a single nucleus is about 200 MeV. The mass of a uranium 235 nucleus is about 4×10^{-25} kg. Prove for yourself that 1 kilogram of uranium 235 would release about 100 million million joules if completely fissioned. This is about 10 million times more energy than is released by burning 1 kilogram of fossil fuel.

Figure 12.2 Binding energy.

QUESTIONS

1   A neutron strikes a $^{235}_{92}U$ nucleus and causes it to fission. Three neutrons are released in this process. One of the fission fragments is the xenon isotope $^{136}_{54}Xe$. How many neutrons and how many protons does the other fission fragment contain?

2   (a) What is meant by binding energy?
    (b) Calculate the binding energy per nucleon of a nucleus of the uranium isotope $^{238}_{92}U$ (mass = 238.05076 u). The mass of a proton = 1.00728 u and the mass of a neutron = 1.00867 u.

3   (a) Explain what is meant by a chain reaction.
    (b) In a certain chain reaction, each fission event releases on average two neutrons that go on to cause further fission after 0.1 seconds. How many nuclei would have been fissioned 1 second after an initial fission event?

Nuclear reactors

A nuclear reactor in normal operation releases energy at a steady rate. This is because the rate of fission events in the reactor is controlled by **control rods** so that each fission event causes exactly one further fission event. The control rods are made of material that absorbs neutrons and they can be moved in or out of the reactor core to keep the number of neutrons in the core constant. Should the fission rate rise, the control rods are pushed into the core a little to mop up surplus neutrons and thus reduce the fission rate.

The mass of uranium 235 must exceed the **critical mass**, which is the least mass for fission to be sustained. If the mass of uranium 235 is less than the critical mass, too many neutrons escape or are absorbed by uranium 238 if this is present. The fuel in a nuclear reactor is enriched uranium contained in sealed fuel rods. Natural uranium is mostly uranium 238 ($^{238}_{92}$U) and it contains less than 1% uranium 235 ($^{235}_{92}$U). Uranium 238 absorbs neutrons without fission. For this reason, the percentage of uranium 235 must be increased to about 2–3% (i.e. enriched); otherwise too many neutrons are absorbed without fission or they escape from the reactor core.

The first atom bomb consisted of two sub-critical hemispheres of pure uranium 235 placed at opposite ends of a hollow cylinder. When the device was detonated, the two hemispheres were forced together into a super-critical mass which then exploded.

Insight

A large hot potato cools more slowly than a small hot potato. This is because a large potato has a smaller surface area relative to its mass than a small potato. Therefore it loses less heat per unit mass than a small potato. The same idea applies to loss of fission neutrons from the uranium fuel in a nuclear reactor. A large mass of nuclear fuel loses fewer fission neutrons per kilogram of fuel than a small mass of nuclear fuel.

THERMAL NUCLEAR REACTORS

The neutrons released when a uranium 235 nucleus fissions in a nuclear reactor need to be slowed down. This is because, when they are released, they move too fast to cause further fission. The slowing down is achieved by surrounding the fuel rods that contain the uranium 235 with a suitable substance to 'moderate' the speed of the neutrons. This **moderator** needs to be composed of light atoms. The neutrons collide with the moderator atoms and transfer kinetic energy to them in the process. This transfer process is most effective if the moderator atoms are as light as possible. The moderator must also be unreactive. For these reasons, water or graphite (which is carbon) is used as the moderator. After many collisions with moderator atoms, the neutrons move slowly enough to cause further fission. They are then referred to as **thermal neutrons**, because their kinetic energies are of the same order of magnitude as the moderator atoms.

The design of one type of thermal nuclear reactor is shown in Figure 12.3. The fuel rods are placed in channels in the moderator, which is a solid graphite core. The control rods are in channels between the fuel rods. The moderator becomes very hot because its atoms absorb kinetic energy from the neutrons, and so carbon dioxide gas as a coolant fluid is pumped through the moderator channels. The entire core is enclosed in a sealed steel vessel, which is connected via inlet and outlet pipes to a heat exchanger. The coolant is pumped round this sealed circuit and used to raise steam in the heat exchanger to drive electricity generators.

FUEL USAGE

The complete fission of a kilogram of uranium 235 releases about 8×10^{13} J of energy (= 80 million million joules). A 1000 MW power station operating at an efficiency of 25% needs 4000 million joules of energy from its fuel every second. Each day, the fuel would therefore need to release about 350 million million joules. About 4 kilograms of uranium 235 would therefore be needed to keep a 1000 MW nuclear reactor operating for a day. If the fuel

2 The moderator slows down the neutrons from each fission event, so they can produce fission of more U-235. Otherwise they are absorbed by U-238 without producing fission.

concrete case

coolant

3 The control rods absorb excess neutrons to ensure only one neutron per fission produces further fission.

steam out

steam out

1 The fuel rods contain enriched uranium, which is 97% U-238 and 3% U-235.

4 Energy released by fission is removed as heat by a coolant, which is pumped through the reactor core. The energy released is colossal: 1kg of U-235 releases more energy than 200 tonnes of coal.

water in

water in

collant pumps

6 The spent fuel is highly radioactive and must be stored for many years after removal from the reactor.

5 The core is in a thick-walled steel vessel encased in concrete to prevent neutrons and radioactive particles from escaping.

Figure 12.3 A thermal nuclear reactor.

for the reactor contained 2% uranium 235, then the mass of fuel needed per day would be about 200 kg (= 0.2 tonnes). In contrast, a 1000 MW oil-fired power station uses about 40 000 tonnes of oil per day.

Chernobyl

In 1986 a nuclear reactor at Chernobyl in the then USSR exploded and scattered radioactive substances into the atmosphere and over the surrounding land. Within days these substances were scattered by the atmosphere over northern and western Europe, including Britain. Upland regions that experienced rainfall in this period were most affected, as the rainwater contained particles of radioactive isotopes. The fuel rods in a nuclear reactor contain many different radioactive isotopes, including uranium 238, plutonium 239 (produced as a result of uranium 238 absorbing neutrons) and neutron-rich fission fragments. The half-lives of these isotopes range from fractions of a second to millions of years. Background count rates increased temporarily in the weeks after the Chernobyl disaster. Many thousands of people in the districts surrounding Chernobyl were moved permanently to other parts of the USSR. Children have been most affected as a result of illnesses such as leukemia caused by radioactive particles entering their bodies. The reactor used steam as the coolant. A subsequent inquiry found that design faults and human error caused a chemical explosion in the reactor during a test when the emergency safety system was switched off. The explosion blew the top off the reactor and radioactive fission products were released into the atmosphere. Many of the emergency workers died subsequently from diseases caused by exposure to radioactivity.

QUESTIONS

4   Explain what is meant by the term 'critical mass' in relation to nuclear fission.

5   Explain the action of (a) the moderator, (b) the coolant in a nuclear reactor.

6 *Use the following data to calculate the mass of fuel used per week in a 1000 MW nuclear reactor operating at an efficiency of 25%. Energy released per kilogram of uranium 235 = 8 × 1013 J; uranium 235 content of the fuel = 2%.*

Energy options for the future

The demand for energy is likely to increase in the next few decades. As we saw in Chapter 6, oil and gas provide more than 60% of the world's energy supplies. However, present known reserves will probably be used up within the next 50 years. Coal provides about a third of the world's energy supplies at present. Known reserves of coal could last up to three centuries at the present rate of usage. About 8% of worldwide energy demand is met by nuclear power stations, which provide between a quarter and a third of the electricity supplies in industrial countries. World reserves of uranium will probably last about 50 years at the present rate of use. Renewable sources of energy, such as hydroelectricity, supply less than 3% of the world's energy demand. When the world's oil and gas reserves are used up, coal could be used instead, but the increased rate of use would shorten the lifetime of the world's coal reserves to little more than a century. In addition, greenhouse gases produced by burning fossil fuels would still be produced at an excessive rate. The energy options for the future are probably limited to the increased use of nuclear power and the development of renewable energy resources on a much larger scale than at present. We looked at different renewable energy resources in Chapter 6. World uranium reserves are unlikely to last beyond the end of the century and nuclear reactors are not in favour politically at present. Below we will look at two other nuclear options which could be developed more.

NUCLEAR FUSION

How does the Sun produce its energy? It radiates energy at a colossal rate of 400 million million million million joules every second. A millionth of the energy radiated by the Sun in 1 second would be enough to meet the world's energy needs for a whole year. In fact, the Sun produces energy in its core as a result of fusing hydrogen nuclei to form helium nuclei. This process started about 5000 million years ago when the Sun was formed and is likely to continue for another 5000 million years before the Sun runs out of hydrogen in its core.

To make hydrogen atoms fuse to form helium atoms, the atoms must overcome the electrostatic force of repulsion between them. To do this, they must collide at speeds over a thousand times faster than they would have in the Earth's atmosphere. The steps in the process of the formation of a helium nucleus from hydrogen nuclei are as follows:

1 *Two hydrogen nuclei colliding at such speeds fuse to form a nucleus of the hydrogen isotope deuterium $^2_1$H*
2 *A high-speed collision between a deuterium nucleus and another hydrogen nucleus causes a nucleus of the helium isotope $^3_2$He to form.*
3 *A collision between two of these helium nuclei causes the formation of a nucleus of the stable helium isotope $^4_2$He and the release of two protons.*

Insight
If the speed is too low, the nuclei repel each other due to their like charge.

The very high speeds necessary for the process to work are achieved in the Sun because the energy released heats its core to more than 10 million degrees. Such high temperature conditions have been achieved briefly in **nuclear fusion reactors**, enabling light nuclei to be fused and release energy in the process. In such conditions, the atoms lose their electrons and the nuclei move about as a 'plasma', without carrying electrons with them. The plasma is contained in a doughnut-shaped tube by means of magnetic fields, but it loses all its energy if it touches the tube. This containment problem is the reason why fusion at present cannot be made to last more than a short time. Thus nuclear fusion reactors remain under development at present and are unlikely to contribute significantly to world energy supplies for at least another 50 years.

FAST BREEDER REACTORS

Natural uranium contains less than 1% of uranium 235. The other 99% is uranium 238. The spent fuel rods from a thermal nuclear reactor contain unused uranium 238 and the plutonium isotope $^{239}_{94}Pu$. Plutonium is an artificial element that is made in a nuclear reactor when uranium 238 nuclei absorb neutrons. Plutonium 239 can be fissioned by fast neutrons. Thus a chain reaction is possible without the need for a moderator. A reactor that uses plutonium 239 as its fuel is called a fast breeder reactor because

- ▶ *the fuel is fissioned by fast neutrons, and*
- ▶ *the reactor can breed its own fuel by allowing neutrons from the plutonium core to be absorbed by a 'blanket' of uranium 238 surrounding the core.*

Thus a fast breeder reactor creates its own fuel from uranium 238, in the form of plutonium 239. The entire content of natural uranium can thus be fissioned, not just the uranium 235 content.

Fast breeder reactors would therefore extend the lifetime of the world's uranium supplies from a century or so to several thousand years. However, the spent fuel would still contain many radioactive isotopes from the fission fragments and the danger of plutonium leakage to the environment would increase as greater quantities of plutonium would be in use. Plutonium 239 has a half-life of 24 000 years and is much more radioactive than uranium 238 or uranium 235. A further danger is that plutonium 239 could accumulate inadvertently in the process of manufacturing the fuel rods; if too much plutonium accumulated, an uncontrollable chain reaction could be started, ending in a nuclear explosion. The experimental fast breeder reactor at Dounreay in Scotland proved that such a reactor works, but it has been decommissioned because of the risk of plutonium leakage. New fast breeder reactors are unlikely to be constructed in the foreseeable future. However, unless renewable resources are developed on a much larger scale, plutonium reactors may be necessary.

QUESTIONS

7   State one advantage and one disadvantage of (a) coal and
(b) uranium as a fuel to provide electricity.

8   (a) What is meant by nuclear fusion?
(b) Give one technical reason why nuclear fusion is difficult to
sustain in a reactor.

9   (a) What type of fuel is used in a fast breeder reactor?
(b) Give one advantage and one disadvantage of the fast breeder
reactor compared with the thermal nuclear reactor.

10 Discuss the range of energy options (including renewable resources
described in Chapter 6) that are likely to be available after the
world's oil and gas reserves are used up.

10 THINGS TO REMEMBER

1 *The* **strong nuclear force** *is an attractive force between the neutrons and protons in the nucleus so as to hold them together.*

2 *The* **binding energy** *of a nucleus is the energy needed to separate a nucleus into separate neutrons and protons.*

3 *The binding energy per nucleon of a nucleus is its binding energy ÷ the number of protons and neutrons it contains. The greater the binding energy per nucleon of a nucleus, the more stable it is.*

4 **Nuclear fission** *is the splitting of a large unstable nucleus into two approximately equal smaller nuclei when a neutron collides with it.*

5 *When a uranium 235 nucleus undergoes fission, several 'fission' neutrons are released which can go on to produce further fission.*

6 *A* **chain reaction** *occurs when one or more fission neutrons from a fission event cause further fission events, which release fission neutrons that go on to cause further fission events, etc.*

7 *In a* **thermal nuclear reactor**, *the fission neutrons are slowed down by the moderator so that they cause further fission. The fission neutrons lose kinetic energy as a result of repeated collisions with the moderator atoms.*

8 **Nuclear fusion** *is when two small nuclei moving at high speed collide with each other and fuse together to form a heavier nucleus.*

9 *Energy is released when nuclear fusion occurs.*

10 *Nuclear fusion occurs in the core of the Sun.*

13

Journey into space

In this chapter you will learn:
- *about Newton's theory of gravity and its limitations*
- *how Einstein predicted that gravity bends space*
- *about the Big Bang theory and the future of the Universe.*

One of the great successes of modern science is the discovery that the Universe originated about 13 000 million years ago in a cataclysmic explosion, the Big Bang, which created space and time. The enormous energy released in the Big Bang created particles and antiparticles which annihilated to produce radiation and excess particles. The entire Universe has been expanding ever since the Big Bang. All the matter in the Universe has formed from the excess particles left after the early stages of the Big Bang. How will the Universe change in future? This is a problem that physicists may well be able to answer soon. In this chapter we will look at the discoveries that have led to our present state of knowledge.

Physics and astronomy have always been at the cutting edge of human endeavour. Newton put the two subjects on a firm mathematical basis when he wrote *The Principia*, published in 1687, in which he set out the mathematical principles of physics. He showed that these principles govern the motion of objects in space as well as objects on the Earth. Newton's laws of motion and of gravity provided the framework for the development of physics for over two centuries, until Einstein produced his theories of relativity. We will look in this chapter at Newton's theory of gravity and Einstein's theory of general relativity which replaced it and which led to predictions, now confirmed, such as the bending of light by gravity, black holes and the expansion of the Universe.

About gravity

From Earth, the planets appear as wandering stars because each planet, as seen against the distant stars, changes its position relative to the Earth as it orbits the Sun. Johann Kepler (1571–1630) devised three laws to describe the motion of the planets, using his own observations and those of his predecessor in Prague, Tycho Brahe (1546–1601), the foremost astronomer of his era.

Kepler's 1st Law: *Each planet moves on an elliptical path round the Sun.*

This law introduced the new idea that the orbits of the planets are in general ellipses, not circles. Figure 13.1(a) shows how to sketch an ellipse. Kepler said that each planet's orbit was an ellipse, with the Sun at one of its focal points.

Kepler's 2nd Law: *The straight line from the Sun to a planet sweeps through equal areas in equal times as the planet moves round the Sun.*

The 2nd Law arose because Kepler knew from his observations that the progress of Mars at its maximum distance from the Sun is 0.8 times slower than its progress at its least distance. He thus worked out that if an imaginary straight line from the Sun to Mars took a certain amount of time to sweep through an angle of 10° at minimum distance, the line would take the same amount of time to sweep through an angle of 8° at maximum distance. He also knew that the least distance from the Sun to the planet Mars is about 0.9 times the maximum distance, and he showed that a sweep of 8° at maximum distance covers an area equal to the area swept for 10° at minimum distance.

(a)

(b)

Times from P₁ to P₂ = Time from P₃ to P₄
if area SP₁P₂ = SP₃P₄

Figure 13.1 (a) Making an ellipse. (b) Kepler's 2nd law.

Kepler's 3rd Law: *The cube of the mean radius of the orbit of a planet is in proportion to the square of the time taken by the planet for each orbit.*

Kepler's 3rd Law required the measurement of the mean radius of orbit r of each planet and the time T each planet takes to orbit the Sun. The results of these measurements are given in the table below. Note that the value of r^3/T^2 is the same for every planet. Thus Kepler's 3rd Law follows: namely, r^3 is proportional to T^2.

	Mercury	Venus	Earth	Mars	Jupiter	Saturn
r in A.U.	0.39	0.72	1.0	1.5	5.2	9.5
T in years	0.24	0.61	1.0	1.9	11.9	29.5
r^3/T^2	1.0	1.0	1.0	1.0	1.0	1.0

1 A.U. = 1 astronomical unit = the mean distance from the Sun to the Earth.

Insight

Unlike Mars, the Earth's orbit is almost perfectly circular. If it were not, we would probably experience an extra summer each year when the Earth is closest to the Sun and an extra winter when it is furthest from the Sun.

NEWTON'S THEORY OF UNIVERSAL GRAVITATION

The reason why the planets move round the Sun remained a mystery until Newton developed the theory of gravity. Newton had already worked out the laws of motion (see pp. 35–42) and he knew that a force must act on each planet to keep it moving round the Sun. He realized that this force must be an attractive force and that it depended on the mass of the planet, the mass of the Sun and the distance from the planet to the Sun. He then deduced that the force of gravity varies with the inverse of the square of the distance from the planet to the centre of the Sun. In other words, at 2 A.U. from the Sun, the force of gravity on an object due to the Sun is a quarter of what it would be on the same object at 1 A.U.; at 3 A.U., the force would be a ninth of what it would be on the same object at 1 A.U.

Newton made this prediction by taking the following steps:

▶ He worked out the theory of motion for any object moving round a circle of radius r at a constant speed v and proved that the force needed is proportional to v^2/r.

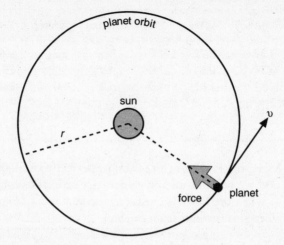

Figure 13.2 Newton's theory of universal gravitation.

▶ *From Kepler's 3rd Law, he knew that r^3/T^2 is constant and as the speed v is equal to $2\pi r/T$, where $2\pi r$ is the circumference of the orbit, he deduced that $v^2 r$ is constant.*

▶ *Newton then showed that as $v^2 r$ is constant, then v^2/r must be proportional to $1/r^2$.*

Thus the force of gravitational attraction between the Sun and a planet is proportional to $1/r^2$.

Newton put forward the theory that the force of gravitational attraction between any two objects is:

1 *proportional to the product of the masses of the objects*
2 *inversely proportional to the square of the distance between their centres of mass.*

He then proved the general result that this force law caused planets to move on elliptical paths round the Sun. He also showed that the inverse square law explained the motion of comets round the Sun, the Moon's orbital motion round the Earth and how the Moon's gravity causes the ocean tides on the Earth. Newton's theory of gravity is described as a universal theory because it explains the motion of objects on the Earth as well as in space. However, a minor problem was discovered in the nineteenth century in connection with the orbit of Mercury, which could not be explained by Newton's theory. The problem remained until Einstein solved it using his general theory of relativity, a theory that drastically altered our understanding of space and time. We shall look at the consequences of this revolutionary theory in the next part of this chapter.

How to lose weight

The pull of gravity on an object on the Moon is one-sixth of the pull of gravity on the same object on the Earth. A 60 kilogram person has a weight of about 600 newtons on the Earth, but only about 100 newtons on the Moon.

- ▶ The Moon's mass is about 1% of the Earth's mass, which would reduce lunar gravity to 1% of Earth's gravity if the Moon had the same radius as the Earth.
- ▶ However, the Moon is one-quarter of the size of the Earth and so its surface gravity is 16 times stronger than if the Earth and Moon were the same size.
- ▶ Hence lunar gravity is 16% (about one-sixth) of Earth's gravity.

EINSTEIN'S NEW UNIVERSE

In his 1905 theory of relativity, Einstein showed that space and time are not separate quantities. By assuming that the speed of light is independent of the motion of the source and of any observer, he predicted such effects as length contraction, time dilation and relativistic mass (see pp. 186–88). In his 1905 theory, he showed that all uniform motion is relative and that absolute uniform motion cannot be detected and is therefore meaningless. In other words, he showed that the laws of physics are the same for all observers moving at constant velocity relative to each other. From 1905, he spent more than ten years developing his ideas on space and time in order to prove the general principle of relativity that the laws of physics are the same for all observers in non-uniform motion as well as in uniform motion. In the course of this work, he proved that accelerated motion and motion due to gravity cannot be distinguished and are therefore equivalent. Thus he reduced gravity to a feature of space and time and went on to prove that gravity bends light as a result. He published his general theory of relativity in 1916 and used it to explain the astronomical puzzle about the elliptical orbit of the planet Mercury, which had confounded scientists ever since the problem was discovered in the previous century. Careful measurements of the position of the perihelion of the orbit had revealed that this position was slowly advancing at a rate of 1.2 hundredths of a degree per century. The perihelion is the nearest point along the orbit to the Sun. Einstein proved that this perihelion advance was caused by the Sun's gravity distorting space near to the Sun.

Einstein also predicted that starlight grazing the Sun would be deflected, an effect he said could be measured by observing stars near the Sun during a solar eclipse. When this prediction was confirmed by British astronomers in 1919, Einstein became a celebrity around the world as 'the person who knew how to bend light'. More significantly, his theory of general relativity was confirmed and he thus proved that absolute space and absolute time do not exist.

Figure 13.3 Mercury's orbit.

Einstein's explanation of Mercury's orbit wasn't accepted until his 'solar eclipse' prediction was confirmed. There is no other known explanation for the deflection of starlight from stars near the Sun during a solar eclipse.

GRAVITATIONAL LENSES AND BLACK HOLES

Further evidence that gravity bends light has been obtained by astronomers in recent years as a result of the observation of distorted images of some distant galaxies. A galaxy is an enormous collection of stars held together by their own gravity. The Sun is part of the Milky Way galaxy, which is reckoned to contain more than a thousand million stars. The space between galaxies is thought to be mostly empty. Distorted images are caused by light from a distant galaxy skimming the edge of an invisible dark galaxy before reaching us. The dark galaxy acts as a **gravitational lens** and bends this light, causing us to see a distorted image of the distant galaxy that emitted the light. Often, the galaxy that emitted the light is hidden completely behind the dark galaxy and so we see its image only as patchy rings of light. The same effect can be observed if a torch bulb is observed through a glass lens that is partly covered by an opaque disc; the light from the torch bulb reaches your eye via the rim of the lens, so you see the rim as a ring of light.

Gravitational lensing provides unexpected evidence in support of Einstein's general theory of relativity. Recent discoveries of **black holes** in space are further evidence in support of Einstein's ideas. Such objects were predicted by Karl Schwarzchild, using the general theory of relativity a few years after Einstein published it. Not even light can escape from a black hole. No object inside a black hole can escape from it. The mass of a black hole is so large that nothing, not even light, can escape from its gravity. Imagine a rocket projected into space from the surface of a planet. The rocket will fall back to the surface if its speed of projection is less than a certain value known as the escape speed. The larger the mass of the planet, the bigger the speed needed to escape from it. Nothing can travel faster than light so if the mass of the planet is sufficiently large, nothing can escape from it.

An object that falls into a black hole is trapped for ever. The **event horizon** of a black hole is its boundary, as any object on or inside the boundary is trapped for ever. A rocket approaching a black hole could pass through the event horizon into the black hole, thus becoming trapped for ever. No signals could be sent from the rocket to observers outside the event horizon because nothing, not even light, can escape from the black hole. Schwarzchild showed that the event horizon of a black hole is a sphere of radius in proportion to its mass. For a black hole of mass equal to the Sun's mass, the radius of the event horizon would be about three kilometres. Such a black hole would pull in nearby stars, becoming even larger in the process.

Insight

A black hole is a cosmic dustbin. Anything put into it can never come out. But when it gains mass, it becomes bigger and so can attract more mass and become even bigger – a cosmic dustbin that devours everything nearby!

Evidence for black holes

Astronomers reckon that so-called **active galaxies,** which spew out jets of matter at colossal energies, each have a black hole at the centre that pulls in and destroys stars. Less active galaxies such as M87 have stars swirling round a black hole at the centre. There could even be a black hole at the centre of the Milky Way, our home galaxy. Fortunately, the Sun is in the outer spiral arms of the galaxy, not at its centre.

A star about 10 or more times larger than the Sun would become a black hole if it collapsed without loss of mass to a ball a hundred thousand times smaller. In fact, such a dramatic collapse, known as a **supernova,** happens at the end of the life cycle of a giant star. The core of such a star collapses when all its light elements have been fused to form heavier elements (see pp. 221–222). The outer layers of the star are thrown off in a massive explosion and the core is compressed so much that a black hole is formed. Such a dramatic end will not happen to the Sun, as it is not massive enough.

QUESTIONS

1 A comet moves on an elliptical path round the Sun, stretching far beyond the Earth's orbit.
   (a) Sketch a diagram to show the orbit of a comet. Mark the position of the Sun and draw a circle to represent the Earth's orbit.
   (b) Explain why a comet takes far more time to return to the inner solar system than to pass through the inner solar system.

2 (a) Jupiter orbits the Sun once every 11.8 years. Use Kepler's 3rd Law to prove that its radius of orbit is 5.2 A.U.
   (b) The mean radius of Saturn's orbit is 9.5 A.U. Use Kepler's 3rd Law to prove that Saturn orbits the Sun once every 29 years.

3 (a) State two reasons why the Moon's surface gravity is weaker than that of the Earth.
   (b) The mass of Mars is 11% of the mass of the Earth. The radius of Mars is 53% of the radius of the Earth. Use this information to show that the surface gravity of Mars is 38% of that of the Earth.

4 (a) On a sketch diagram showing the Earth and the Sun, sketch the path of a radio signal sent from a spacecraft to the Earth when their positions are such that the signal grazes the Sun.
   (b) With the aid of a sketch diagram, describe how a galaxy can act as a gravitational lens.

5 (a) State one characteristic feature of a black hole.
   (b) Describe what is meant by the event horizon of a black hole.

The expanding Universe

By the end of the twentieth century, astronomers had worked out that the Universe is expanding. The distant galaxies are moving away from us as the galaxies move apart due to the expansion of the Universe. However, two centuries earlier, most scientists believed the Universe to be infinite in space and in time, the same as it always had been. This view was challenged in 1826 by Heinrich Olbers, who proved mathematically that the sky would be permanently bright if there was an infinite number of stars. So Olbers concluded that the Universe is finite and he reckoned that it must be expanding; otherwise, gravitational attraction would cause it to collapse. A century later, Einstein used his general theory of relativity to show that a static finite Universe is possible, but he needed to include a cosmic repulsion force to stop the Universe collapsing. The question of whether the Universe is static or in a state of collapse or expansion was settled by the astronomer Edwin Hubble in 1929 when he published the results of his survey of about two dozen galaxies at known distances within about 6 million light years of the Milky Way galaxy. Hubble discovered that the spectrum of light from each galaxy is shifted towards the red part of the spectrum. This shift is known as a **red shift** and it happens because the light waves from a receding light source are lengthened as it moves away.

Insight

The frequency of the sound waves from an emergency vehicle siren rises and then drops as the vehicle races towards and then past you. This is because the motion of the vehicle causes the sound waves to 'bunch up' in front of the vehicle and stretch out behind it.

Hubble knew that the increase of the wavelength of light due to a red shift is in proportion to the speed at which the light source moves away (i.e. recedes). He measured the wavelength change for the light from each galaxy and used the measurements to calculate the speed of recession of each galaxy. His results showed that the speed of recession of a galaxy is in proportion to its distance from the Sun. This discovery is known as **Hubble's Law.**

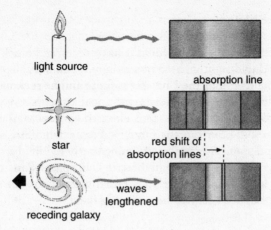

light source

absorption line

star

red shift of
absorption lines

waves
lengthened

receding galaxy

Figure 13.4 Red shift.

Why should all these galaxies be rushing away from us? The Earth
has no special place in the Universe. Hubble realized that the
Universe is expanding so that

▶ *every galaxy is moving away from every other galaxy*
▶ *the further a galaxy is from us, the greater its speed of
recession.*

Make a model of the expanding Universe

Mark a rubber band with spots along its length at equal intervals.
The spots represent galaxies. Observe how the spots move away
from each other when the band is stretched slowly. Two spots
initially near to each other move away from each other much
more slowly than two spots initially far from each other.

THE BIG BANG

If the Universe is expanding, could it have once been much smaller? In fact, the Universe originated in a massive primordial explosion, known as the **Big Bang**, and has been expanding ever since. The Big Bang theory was opposed by some scientists who believed the Universe to be in a steady state, expanding because of matter pouring into it at points in space referred to as 'white holes'. In 1965 the dispute was settled unambiguously by the discovery of microwave radiation from all directions in space. This cosmic microwave background radiation is radiation that has been travelling though space ever since the Big Bang.

More accurate measurements of the red shift and distances to galaxies further and further away continue to support Hubble's Law. These measurements indicate that the speed of a galaxy is 20 kilometres per second for every million light years. Thus a galaxy that is moving away from us at a speed of 20 000 kilometres per second must be 1000 million light years distant. Since no object can travel faster than the speed of light, which is 300 000 kilometres per second, then the most distant galaxies cannot be further away than about 15 000 million light years. Thus the Big Bang cannot have taken place more than 15 000 million years ago, which is therefore the age of the Universe. Taking account of gravity, because it holds back the expansion of the Universe, gives an age of about 13 000 million years. In the next section, we will meet quarks and antiquarks again when we consider the chain of events in the early stages of the Big Bang.

THE FUTURE OF THE UNIVERSE

The theory of the expanding Universe without a cosmic repulsion force had been worked out by Alexander Friedmann in Russia in 1922, using Einstein's general theory of relativity. Einstein described his earlier idea of a cosmic repulsion force as his 'greatest blunder'. However, recent observations indicate that the expansion is accelerating, which suggests a force of repulsion could be driving the galaxies away from each other faster and faster.

Einstein's general theory of relativity predicts that the future of the Universe depends on the average density of matter in it, which is not known at present.

▶ *If the density is too great, the Universe will stop expanding and then start to contract, ending in the* **Big Crunch**.
▶ *If the density is too small, the Universe will expand for ever, a scenario known as the* **Big Yawn**.
▶ *If the density is equal to a certain value, the Universe will expand at an ever-decreasing rate without ever reversing.*

Recent observational evidence from accurate measurements of distant supernovas suggests that the expansion of the Universe is accelerating. The amount of known matter in the Universe is insufficient to prevent a runaway expansion. Some form of invisible **dark energy** not yet known about might be making the galaxies accelerate as they move away from each other.

THE EARLY UNIVERSE

The cause of the Big Bang is not yet known. What is known is that a vast amount of energy was suddenly released, creating space and time in the process and causing space to expand from nothing to the size of a football in less than a billionth of a billionth of a billionth of a microsecond.

▶ *By this stage, the temperature had fallen to about a million million degrees, allowing quarks, antiquarks, electrons and positrons to form from radiation.*
▶ *After about a minute, the temperature had fallen to about a thousand million degrees. The antiquarks were annihilated by the quarks and the positrons were annihilated by the electrons. The remaining quarks formed neutrons and protons. At this stage, the Universe had expanded to a size bigger than the Solar System.*
▶ *The neutrons and protons joined together to form nuclei as the Universe cooled further. After about 100 000 years, the temperature had dropped to a few thousand degrees.*

Atoms formed as nuclei were able to hold on to electrons. The Universe became transparent at this stage because atoms absorb light far less than ions do. Thus the dark age of the Universe began.

Insight

The spots on an untuned TV set are caused by cosmic microwave background radiation picked up by the TV aerial. Such radiation has been travelling through space ever since the Universe became transparent.

▶ *After a few billion years, galaxies began to form and move away from each other as the Universe continued to expand. Astronomers do not yet know if stars gathered to form galaxies or if stars formed inside galaxies.*

Thus the Universe, which is matter on the largest possible scale, has developed as a result of the properties of matter and radiation on the smallest possible scale. Many questions remain to be answered, including how gravity works and why matter and antimatter did not annihilate each other completely in the early stages of the Big Bang. Clearly, this did not happen; otherwise we would not exist.

QUESTIONS

6 (a) Explain what is meant by the 'red shift'.
  (b) What did Hubble discover about the link between the red shift and the distance to a galaxy?
  (c) What conclusion about the Universe was drawn from Hubble's discovery?

7 A galaxy at a distance of 100 million light years was found to be moving away from us at a speed of 2000 kilometres per second. Use this information to estimate:
  (a) the distance to a galaxy moving away from us at a speed of 8000 km/s
  (b) the speed of recession of a galaxy at a distance of 4000 million light years.

8 (a) What is the Big Bang theory of the Universe?
   (b) What discovery caused the 'steady state' theory of the Universe
       to be rejected in favour of the Big Bang theory?

9 Put the following events in order of time of formation,
   commencing with the Big Bang:
       Big Bang
       Atoms formed
       Galaxies formed
       Neutrons and protons formed
       Quarks and antiquarks formed

10 (a) What was the approximate age of the Universe when
       (i) neutrons and protons formed, (ii) it became transparent?
   (b) What is thought to be the most likely future of the Universe?

10 THINGS TO REMEMBER

1 **Newton's theory of gravitation** *tells us that there is an attractive force between any two objects.*

2 *The force of gravity between two objects is inversely proportional to the square of the distance between their centres of mass.*

3 *Newton's theory of gravitation explains many observations, including the motion of the planets round the Sun.*

4 **Einstein's theory of gravity** *explains effects that cannot be explained using Newton's theory, such as black holes and the bending of light by gravity.*

5 *A* **black hole** *is an object from which nothing, not even light, can escape.*

6 *The* **event horizon** *of a black hole is the boundary surrounding it. Nothing inside the boundary can escape.*

7 *The* **red shift** *of a galaxy is the increase of wavelength of light from the galaxy due to the galaxy's receding motion.*

8 **Hubble's Law** *states that distant galaxies are moving away from us at speeds in proportion to their distance from us.*

9 *The Universe originated in a massive explosion referred to as the* **Big Bang**.

10 *The* **expansion of the Universe** *is thought to be accelerating.*

14

A challenging future

In this chapter you will learn:
* *about the present frontiers of physics*
* *why physics can help us avoid catastrophe*
* *about the challenges of physics in the future.*

In this final chapter we will look at some of the problems that
are at the frontiers of physics at present. The topics chosen are
of necessity described briefly and other topics could have been
chosen instead. Nevertheless, the aim here is to outline some of
the problems of physics now and to consider the immense benefits
that could follow if they were solved. Of course, no one can predict
how research into these problems may or may not develop or
where new problems and opportunities might arise. At the end of
the nineteenth century, many physicists reckoned that little else
remained to be discovered except some apparently minor problems
to do with radioactivity and light. Ten years into the twentieth
century, physics had been revolutionized by the quantum theory
and the theory of relativity. In the last few decades, more scientific
knowledge has been gained than in the entire previous history of
the human race. Developments in physics in the next few decades
may or may not be as revolutionary as a century ago, but the work
of physicists will undoubtedly continue to provide major benefits to
us all.

The frontiers of physics

UNDERSTANDING ENERGY

We talk about energy, we measure it, we have government departments to control it, and yet we still do not fully understand it. We cannot see it and yet we know when an object has energy, because we can then make it do some work. Einstein worked out that the mass of an object increases as its speed increases, as outlined on p. 187. In effect, when the speed of an object approaches the speed of light, the mass of the object increases and its motion becomes more difficult to change. Einstein's equation $E = mc^2$ tells us how much the mass changes for a certain amount of energy transferred. How energy transfer causes a change of mass is not yet known, even though Einstein worked out the scale of transfer a century ago. Experiments in high-energy physics at laboratories such as CERN, the European Centre for Nuclear Research at Geneva, might provide the answer in the next few years. These experiments are designed to find out why particles such as electrons, protons and neutrons possess mass. An answer to this question could lead on to finding out how particles with mass gain mass when made to travel faster. The mechanism of $E = mc^2$ would then be uncovered, providing a deeper understanding of nuclear fission and fusion.

ENERGY FOR EVERYONE

We all use energy from fuels, especially people in wealthy countries. The rate of use of global energy is about 10 million million watts (i.e. joules per second). Oil, gas and uranium reserves are unlikely to last into the twenty-second century. Coal reserves might perhaps last several centuries. If everyone on the planet were able to use energy at the same rate as in wealthy countries, the rate of global usage would be five times greater. World population is rising fast as well and so fuel reserves are likely to become scarcer and more expensive. Wars and conflicts could follow if this happens. New discoveries of fuel reserves will undoubtedly be made, but such discoveries are unlikely to bridge the gap between growing demand and declining reserves.

In Chapters 6 and 12 we looked at the options for the future in terms of more renewable energy sources and more nuclear power plants. Renewable energy sources using current technology could make a significant contribution, but not enough to plug the 'demand gap' at present levels of investment. If fuel prices increase, there is likely to be greater investment in renewable energy sources, but the rise in living standards of the past century might slow down or stop or even reverse if investment is too slow. New discoveries in science could close the demand gap. For example, hydrogen gas could be used as the fuel for vehicles; hydrogen could be produced and collected from seawater by electrolysis. Solar cells could provide the electricity for this process. At present, solar cells are expensive and inefficient. Research into better and cheaper solar cells could result in the 'hydrogen' economy replacing the 'petrol' economy. In addition, such research could produce high-voltage solar cells capable of producing much more power and voltage than the present generation of solar cells. Each square metre of a solar panel facing the Sun could absorb as much as 1400 watts of solar power. Even on a dull day, solar radiation could generate 100 watts of electrical power per square metre. To meet global energy demand, less than 0.02% of the surface of the planet (i.e. a total area less than the Sahara desert) would need to be covered in solar panels. No nuclear waste, no greenhouse gases, no oil pollution, no acid rain!

NEW MATERIALS

The benefits of new materials reach us all in many ways, for example through lighter vehicles, better thermal insulation, more reliable artificial joints and improved communications. The explosive growth of the internet is the result of the invention of the transistor in the mid-twentieth century and the subsequent development of integrated circuits on silicon chips in the 1970s. Almost everything we do has been affected by internet and multimedia technologies. Further developments in communications such as cable TV have taken place as a result of research into flexible transparent materials used to make optical fibres. Such fibres are capable of carrying much more information than copper wires.

Another potential revolution awaits the discovery of room-temperature superconductors. At present, superconductors, which are electrical conductors with zero resistance, need to be cooled to very low temperatures. When the temperature of a superconductor is raised, its resistance suddenly returns once it reaches a certain temperature referred to as its critical temperature. The highest known critical temperature is not much more than about 130 K, well below room temperature, which is about 290 K. The benefits of room temperature superconductors would include much more efficient transmission of electrical power across long distances, lighter electric motors, economic magnetic levitation transport systems, cheaper medical scanners and faster computers. Superconducting cables could be used to transfer electrical power long distances from solar power stations in remote areas to populated areas. Research into superconductivity continues as scientists attempt to find out if room temperature superconductivity is possible.

THE WORLD WIDE WEB

The internet is a communications network which was devised by the US government in the Cold War to ensure that an enemy country could not knock out vital communications by means of a nuclear strike. A network of powerful computers in different locations were linked to each other to provide many routes between any two computers. An internet message from one computer to another is chopped into a sequence of packets and each packet is sent independently to the same destination by any route. At the destination, the packets are put together to recreate the original information. The individual packets travel via different routes to the same destination. The internet now consists of a permanent network of internet service providers (ISPs) connected to the telephone system. The World Wide Web (WWW) was developed in the 1990s by physicists at CERN, to enable physicists in other countries to gain access to scientific papers stored electronically via the internet. Within a few years, WWW revolutionized commerce, industry, entertainment and communications. Few people recognized the immense potential of the World Wide Web at the time it was set up.

An unpredictable future

BLUE SKIES RESEARCH

No one can tell the effects of a new discovery. Michael Faraday could not have predicted that his discovery of electromagnetic induction (see p. 116) would develop into a worldwide industry and provide energy to consumers on demand. Joseph Thomson, who discovered the electron (see p. 194), saw the development of the electronics industry within his own lifetime, but even he would have been amazed at the wide range of electronics applications now. Lord Rutherford's investigations on the structure of the atom (see p. 192) led to the discovery of the nucleus, yet even Rutherford at that time could not foresee nuclear power and the development of the nuclear industry. Scientific research that might seem obscure or without immediate goals, often referred to as **blue skies research**, has on many occasions produced revolutionary discoveries.

Scientific research driven by precise aims and targets is necessary in companies with expensive equipment where investors expect dividends. Such **applied research** often leads to important discoveries and developments, such as integrated circuits, providing huge returns for investors and inventors. However, blue skies research can revolutionize our understanding of nature, as well as developing the essential underpinning knowledge from which revolutionary inventions and developments often follow. A flavour of blue skies research in two areas of physics is outlined below.

Low-temperature physics is a very active research field as scientists attain lower and lower temperatures, no more than a fraction of a degree above the absolute zero of temperature. The properties of materials at such very low temperatures provide a rich area of research. **Superconductors** were discovered as a result of measuring the electrical resistance of different conductors at such low temperatures. **Superfluidity**, the complete absence of viscosity in a fluid, was discovered when liquid helium was observed emptying itself out of its glass container. Further investigations showed that this fluid lost its viscosity (i.e. its resistance to flow) suddenly at 2.2 K, when cooled from above to below this temperature. It also became more than a million times more effective as a heat conductor below this temperature. Since this discovery, superfluidity has also been discovered in a few other fluids. The explanation of superfluidity lies in the sudden change of behaviour of the helium atoms when cooled below 2.2 K. Instead of moving about at random as they do above 2.2 K, the atoms link together in a single quantum state at or below 2.2 K. Superfluidity is an example of quantum behaviour on a scale large enough to see directly, and so it provides a deeper understanding of quantum theory. In addition, understanding superfluidity has helped scientists to understand superconductivity in metals.

Chaos theory is a recent arrival in science and has opened up many new and interesting applications. The ideas of chaos theory have been applied in such diverse fields as fluid flow, population dynamics in biology, medical research in connection with heart attacks, share prices on the finance markets and asteroid impacts. Chaos theory stems from computer modelling of **non-linear** systems. Some measurable quantity (or quantities) is used to specify the state of a system at regular intervals. If the value of the quantity at some interval does not maintain a constant relationship to the value in the previous interval, the system is described as non-linear. A non-linear system can behave chaotically in certain circumstances, whereas a linear system (where the proportionality is maintained) is reliable and predictable.

To see how the ideas can be applied, consider the population dynamics of animals on a small island. Suppose the island is populated only by a certain animal species, which breeds once each year and which lives off the natural vegetation on the island. The number of animals each year depends on the number the previous year. Too many animals would reduce the vegetation and hence the food supply, cutting the number of animals that survive; too few animals would breed too little. Suppose the island can support a maximum of 100 animals each year. The population change each year depends on breeding, which is proportional to the population P in the previous year, and on the food supply, which is proportional to (100 − P). In other words, the population in a given year depends on how far the previous year's population was from 0 and 100. As a result, the population may change dramatically from one year to the next. Extinction would occur if P reached 100 in any year, as there would be no food supply left.

GLOBAL CATASTROPHE

Many scientists think that **global warming** is the greatest danger we face. The human race on planet Earth is not unlike an animal population on an island. The Earth's atmosphere acts like a greenhouse, in that it traps sufficient heat radiation from the Sun to keep us warm. However, as explained in Chapter 6, we are using increasing amounts of fossil fuel as living standards rise across the world. As a result, the amount of carbon dioxide gas and other so-called 'greenhouse gases' in the atmosphere has increased dramatically over the past 50 years or so. Most scientists reckon that more and more dangerous global warming is taking place due to the increasing amount of greenhouse gases in the atmosphere. Such global warming could have catastrophic effects, including:

▶ *melting the polar icecaps, which would cause raised sea levels everywhere and force mass migration from flooded land areas*
▶ *changing the Earth's climate, which would affect agricultural production and food supplies in many countries and cause more extreme weather events.*

Many scientists are calling for urgent action to reduce greenhouse gases drastically in order to stop global warming. Governments worldwide need to agree and implement a wide range of measures, including:

▶ *capturing and storing carbon dioxide emissions from electricity power stations (as yet the effectiveness of this is unproven)*
▶ *replacing fossil fuel power stations with new nuclear power stations and renewable energy generators such as geothermal,*

tidal and hydroelectric power stations, wind turbines and solar power installations

▶ *making the machines, engines and vehicles we use more efficient and reducing heat losses from buildings in winter*

▶ *improving public transport and replacing petrol vehicles with electric only or hybrid electric/petrol vehicles.*

An **asteroid impact** by a large asteroid could wipe out the human race just as dinosaurs were wiped out about 60 million years ago. Pockets of survivors might be forced to live as our ancestors did, subjected to the perils of nature. The asteroids are very large chunks of rock in orbit about the Sun, mostly between Mars and Jupiter. In addition to the Sun's gravity pulling on an asteroid, Jupiter can affect an asteroid as it is the largest planet of the Solar System. The path of an asteroid therefore depends on its distance from the Sun and its distance from Jupiter. Using chaos theory, astronomers reckon that there is a very small possibility of an asteroid behaving chaotically and colliding with the Earth. No doubt other solar systems have the same problems. Any Earth-like planet in any other solar system would be subject to asteroid impacts from asteroids in its own solar system. The probability of intelligent life elsewhere in the Milky Way galaxy is therefore even more remote if it can be wiped out by asteroid impacts. Perhaps the human race is the only intelligent life form in the galaxy at the present time. Astronomers and scientists are becoming increasingly aware of the need to watch out for collisions with errant asteroids and other large objects from space. The US government has set up an observatory to search for and track asteroids. However, advanced warnings of an asteroid impact would not shield the Earth from an impact. An international defence system needs to be set up to safely deflect or destroy any asteroid or other large object from space found to be on course to collide with the Earth. Governments would need to provide the funds, perhaps recharging people in the wealthy nations through a small annual surcharge on home insurance policies. This might seem far-fetched, but meteorite impacts are not. A large meteorite impact created the Grand Canyon in Arizona. Such an impact on a densely populated area would devastate it. An asteroid impact would devastate entire continents and throw millions of tons of dust into the atmosphere. The sky would be dark for many years and the surface would freeze. Food supplies would dwindle and transport would cease. An international defence system seems a small price to pay in comparison.

FROM 'WHAT IF?' TO 'WOW!'

Physics has always been about big ideas and big discoveries that have affected everyone. Yet most big ideas or discoveries in physics start out with 'What if?' thoughts. Intuition, creativity, imagination, speculation and persistence are the sort of characteristics that are needed to think up new ideas. Fifty years ago, Francis Crick and James Watson successfully modelled the structure of DNA in a Cambridge physics laboratory and so started the scientific world on a trail which has led to the successful completion of the human genome project. Crick and Watson used knowledge about DNA discovered from X-ray diffraction experiments to work out the double helix structure of the DNA molecule. The human genome, the sequence of the four bases that link together in a very long chain to form the DNA molecule, has now been discovered as a result of scientists using electricity to control charged fragments of DNA and using lasers and computers to measure the results. The benefits of genetic engineering are already being felt, for example by sufferers of cystic fibrosis who can now lead normal lives outside the shadow of crippling illness and early death. Such developments will undoubtedly bring many more benefits to many people. The physicists who discovered X-rays, lasers and new electronic devices leading to modern computers could scarcely have imagined how their ideas and discoveries have been utilized by other scientists.

Who knows what can emerge from an idea? At the time of writing, physicists continue to struggle to reconcile quantum theory and relativity theory, the two great theories of modern science. Gravity, which emerges from relativity theory, lies outside the standard model of quarks and leptons, which is rooted in quantum theory. Can these two great theories be brought together? Some physicists think that a combined theory is possible through a mathematical framework referred to as **string theory**. However, others seek a different solution by challenging such basic tenets as the constancy of the speed of light and the principle of conservation of energy. Whatever the solution is, if one is found, the history of physics tells us that new problems will probably then emerge, which will call for further new ideas and fresh insight. The concept of 'dark energy', mentioned on p. 239, to account for the acceleration of the expansion of the Universe, suggests that energy has further properties that we do not know about yet.

SCIENCE AND SOCIETY

Modern science has provided us with the means of finding out about nature on a scale from quarks to quasars. Many more scientific discoveries undoubtedly lie ahead. The observations and experiments carried out by scientists provide essential guidance for the theories of science. Before the Scientific Age was ushered in by Galileo, few dared to challenge the medieval Church on the theory that the Earth must be at the centre of the Universe. The freedoms we now take for granted, such as human rights and democracy, would perhaps not have been won if the struggle for intellectual freedom, the hallmark of modern science since Galileo, had been lost. Perhaps the real long-term challenge for the future is to extend the immense benefits of science to the entire human race and to find out how to sustain these benefits indefinitely. Scientific knowledge is vital for a better future. I hope you have enjoyed using this book and that you will take your physics studies further.

Glossary

absolute zero The lowest possible temperature (= $-273°C$).

acceleration Rate of change of velocity.

atom The smallest particle of an element characteristic of the element. Every atom contains a nucleus (which consists of protons and neutrons, except in the case of the hydrogen atom which has a single proton as its nucleus) surrounded by electrons.

binding energy Energy needed to separate a nucleus into its constituent neutrons and protons.

bond Name for any type of force that holds two particles together.

carbon footprint The mass of carbon dioxide released into the atmosphere per year as a result of the energy you use to maintain your lifestyle.

charge There are two types of charge, referred to as positive and negative. Particles that possess the same type of charge repel each other. Particles that possess opposite types of charge attract each other. Charge is quantized in whole-number multiples of e, the charge of the electron.

density Mass per unit volume of a substance.

diffraction Spreading of waves after passing through a gap or round an obstacle.

diffusion Gradual spread of randomly moving particles in a substance to a uniform distribution.

diode Electronic component that allows current to pass through it in one direction only.

elasticity The physical property of a substance that enables it to regain its shape after being distorted.

electromagnetic spectrum The spectrum of electromagnetic waves. The electromagnetic spectrum comprises gamma rays and X-rays, ultraviolet radiation, visible light, infrared radiation, microwaves, radio waves. All electromagnetic waves travel at a speed of 300 000 km/s through space.

electron A negatively charged particle that is in every atom, moving round the nucleus. The charge of the electron e is equal to 1.6×10^{-19} C.

electron volt(eV) 1.6×10^{-19} J, defined as the work done when an electron moves through a p.d. of 1 volt. 1 MeV = 1.6×10^{-13} J.

energy The capacity to change the motion of an object.

force Any interaction that can change the motion of an object.

frequency The number of cycles of oscillation of an oscillating object, each cycle being from one extreme to the opposite extreme and back.

gradient The gradient of a straight line on a graph is the change of the quantity plotted on the y-axis/corresponding change of the quantity plotted on the x-axis.

grains Crystalline structure in a metal.

gravitation There is a force of gravitational attraction between any two masses. The force is inversely proportional to the square of the distance between the centres of the masses.

greenhouse gases gases in the atmosphere that contribute to global warming.

half-life The time taken for half the number of nuclei of a radioactive isotope to disintegrate.

interference Where two waves pass through each other, reinforcement occurs where crests meet or where troughs meet. Cancellation occurs where a crest meets a trough.

internal energy The energy of an object regardless of its state of motion or its position.

ion A charged atom. An uncharged atom contains an equal number of electrons and protons. Removal of an electron makes the atom into a positive ion. Addition of an electron makes the atom into a negative ion.

ionising radiation radiation such as X-rays and alpha, beta and gamma radiation that creates ions in substances it passes through.

isotope The isotopes of an element are forms of the element which each have the same number of protons but a different number of neutrons in each nucleus.

kinetic energy The energy of an object due to its motion.

lepton Elementary particles such as the electron and the positron that are not quarks.

mass A measure of the quantity of matter in an object.

molecule Two or more atoms joined together.

momentum The product of the mass and the velocity of an object.

neutron An uncharged particle slightly heavier than the proton. Every atom contains a nucleus which is composed of one or more protons and neutrons.

nuclear fission When a large nucleus splits into two fragments.

nuclear fusion When light nuclei are fused together.

photon Light is composed of photons. Each photon is a wave packet of electromagnetic energy. The energy of a photon of frequency f is equal to hf, where h is the Planck constant.

potential energy The energy of an object due to its position.

power Rate of transfer of energy.

pressure Force per unit area acting at right angles to a surface.

proton A positively charged particle and the nucleus of the lightest atom, the hydrogen atom, is a single proton.

quarks Particles that combine in threes to form neutrons or protons.

red shift The increase of the wavelength of light from a receding object, due to its receding motion. This increase causes a shift in the line spectrum of the light from the object towards the red part of the spectrum.

refractive index *(n)* The speed of light in air/the speed of light in the substance.

resistance Voltage per unit current needed to make electricity flow.

scalar A non-directional physical quantity.

speed of light Distance travelled per second by light. The symbol c denotes the speed of light in free space. $c = 300\,000$ km/s.

superconductor An object with zero electrical resistance.

superfluid A fluid that has zero viscosity.

U-value Heat flow per square metre passing through a wall or panel, etc., when the temperature difference across the object is 1°C.

vector A physical quantity that has magnitude and direction, e.g. force.

velocity Rate of change of distance in a given direction.

viscosity A measure of flow resistance in a fluid.

voltage Power per unit current delivered between two points in a circuit.

wavelength The least distance along a wave between two crests.

weight The force of gravity on an object.

work Work is done when a force moves its point of application in the direction of the force.

Taking it further

Develop your interest in physics further by reading one or more of the following books or by visiting some relevant websites listed below.

Further reading

For those who wish to study physics at a higher level
New Understanding Physics for Advanced Level (4th edition) by Jim Breithaupt (ISBN 0-748704314-6), published by Nelson Thornes. This is an all-embracing comprehensive physics textbook written for UK AS and A-level physics courses, including options in astrophysics, electronics, materials science and medical physics. It also covers the requirements of the International Baccalaureate Physics course. It is written to take students from GCSE physics to university physics entrance level.

Foundations of Physics (3rd edition) by Jim Breithaupt (ISBN 0-333-73302-9), published by Palgrave. This book is intended for students who are following a college physics access course or a university foundation physics course. It is also suitable for first-year students on degree programmes in which physics is a subsidiary subject. The book covers the subject from GCSE level to university physics foundation level.

For those who wish to consolidate and widen their knowledge of physics
Key Science Physics (3rd edition) by Jim Breithaupt (ISBN 0-7487-6243-4), published by Nelson Thornes. This is a popular school physics textbook written for UK and international GCSE physics courses. It covers the subject at a slightly higher level than *Understand Physics*.

AQA GCSE Physics by Jim Breithaupt (ISBN 0-7487-9647-9), published by Nelson Thornes. This school physics book is written for students studying for the AQA GCSE Physics specification or for the physics units in the AQA Science and Additional Science specifications.

Einstein – a beginner's guide by Jim Breithaupt (ISBN 0-340-78043-6), published by Hodder & Stoughton Educational. The book presents Einstein's work in historical context, taking the reader through the theory of relativity and its implications from black holes to time travel.

Websites

There are so many physics and astronomy websites you can visit that where to start can be perplexing. News of physics developments can be obtained from the UK or American Institute of Physics websites, listed below. Some further websites for physics organizations are also listed.

> *www.iop.org for the UK Institute of Physics*
> *www.aip.org for the American Institute of Physics*
> *www.stfc.ac.uk for the UK Science and Technology Facilities Council (which includes the Particle Physics and Astronomy Research Council)*
> *http://public.web.cern.ch for the European Organization for Nuclear Research*
> *www.spacecentre.co.uk for the UK National Space Centre, Leicester*
> *www.nasa.gov for the US National Aeronautics and Space Agency (NASA)*
> *www.esa.int for the European Space Agency*

Answers to chapter questions

ANSWERS TO CHAPTER 1 QUESTIONS

1 (a) 1.5×10^4 m (b) 4.5×10^{-5} m (c) 6.5×10^8 kg (d) 3.5×10^{-9} kg
2 (a) 7.5×10^{-2} m (b) 1.59 m (c) 5.6×10^7 m (d) 6.5×10^{-2} kg
    (e) 2.7×10^{-8} kg
3 (a) 10^3 (b) 10^6
4 (a) 12 500 kg (b) 0.005 m^3 (c) 3000 kg/m^3
5 (a) 0.001 m^3 (b) 800 kg/m^3
6 (b) 3.12 kg
7 72 kg
8 6000 kg
9 0.93 g/cm^3
10 7500 kg/m^3

ANSWERS TO CHAPTER 2 QUESTIONS

1 (a) 720 m (b) 4.5×10^5 m (c) 1.8×10^4 m
2 (a) 67 s (b) 25 s (c) 1667 s = 27 min 47 s
3 (a) 12.5 m/s (b) 10 m/s (c) 133 m/s
4 (a) 13 m/s (b) 310 m/s (c) 28 m/s (d) 200 m/s
5 (a) 72 000 m (b) speed of 2nd car = 72 000/3600 = 20 m/s
6 (a) 10 m/s (b) 200 m
7 (a) 0.25 m/s (b) 0.50 m/s (c) 0.0625 m/s^2
8 (a) 60 m/s (b) 2400 m (c) 3.0 m/s^2
9 (a) 15 m/s (b) 11 m
10 (a) 4 m/s^2 (b) 5000 m (d) 7000 m (= 5000 m + an additional 2000 m)

ANSWERS TO CHAPTER 3 QUESTIONS

2 (a)(i) 19.6 N (ii) 39.2 N
3 (a)(i) 8 kg (ii) 103 N (b) 103 N
4 (a)(i) 8 N (ii) 24 N (b)(i) 5 m/s^2 (ii) 4 m/s^2
5 (a) 2 m/s^2 (b) 1600 N
9 (b) 100 N

ANSWERS TO CHAPTER 4 QUESTIONS

2 (a) 2000 J (b) 2000 J
3 (a) 900 000 J (b) 9000 seconds
4 (a) 392 N (b) 1176 J (c) 47 W
5 (a) 2425 J
6 (a) 1000 J (b) 5000 J (c) 20%
7 (a) 550 kJ (b) 0.55
8 (a) 1000 J (b) 50 W
9 (a)(i) 3000 J (ii) 900 kJ (iii) 40 J (b)(i) 392 J (ii) 24 J (iii) 1250 J
10 (a) 8820 J (b) 3750 J

ANSWERS TO CHAPTER 5 QUESTIONS

1 (a)(i) 0°C, 273 K (b)(i) 100°C (ii) 373 K (c) –273°C, 0 K
3 (a) No 1: roof = 1200 W, window = 500 W, wall = 1800 W;
      total = 3500 W
   (b) No 3: roof = 300 W, window = 300 W, wall = 600 W;
      total = 1200 W
   (c) No 1 = £2.52 No 3 = £0.86 Difference = £1.66
4 (a) 567 kJ (b) 32 kJ (c) 599 kJ
5 (a) 25°C (b)(i) 21.6 kJ (ii) 12°C
6 (a) 3.4 MJ (b) 18.8 minutes
7 (a) 1.15 MJ (b) 3.4 kg
8 (a)(i) 840 kJ (ii) 53 kJ (b) 71 W
9 (a)(i) 13 kJ (ii) 67 kJ (b) 67 W
10 (a) 180 kJ (b) 0.078 kg

ANSWERS TO CHAPTER 6 QUESTIONS

2 (a) 0.54 (b) 0.51
3 (a) 10 000 MW approx (b) 45 000 MW approx
5 (a) 6 (b) 2500 (c) 100 km

ANSWERS TO CHAPTER 7 QUESTIONS

1 negative, gains, loses
2 (a) the coulomb
   (b)(i) 0.30 C (ii) 9 C

3 (a) 12 hours (b) 36 W
4 (a) 15 V (b) 37.5 W
6 (b)(i) 0.5 A (b) 24 Ω
9 (b)(i) 1:12 (ii)(1) 36 A (2) 3 A
10 5 A
12 (a) £15.50 (b) 96 units, 68%

ANSWERS TO CHAPTER 8 QUESTIONS

1 1.20 m
2 See p. 129
3 See Figure 8.1 (b)
4 (a) Refraction is the change of direction of a light ray when it passes from one transparent substance or air into another transparent substance. (b) 35°
5 See Figure 8.2(b)
6 (a) Light consists of tiny particles called corpuscles.
(b) The speed of light in water was found to be less than in air, as predicted by wave theory.
9 (a) See summary p. 141 (b) Infrared radiation, radio waves, microwaves
10 (a) Red (b) 500 million million hertz

ANSWERS TO CHAPTER 9 QUESTIONS

1 (a)(i) negative (ii) positive (b)(i) 16 atomic mass units (ii) +8 e, where e is the magnitude of the charge of the electron (iii) +e
2 (a) molecule, atoms, atom (b) isotope, protons, neutrons (c) proton, electron
4 (a) a covalent bond (b) an ionic bond
6 (a) crystalline (b) polymer (c) crystalline (d) amorphous (e) polymer
7 (a) a good electric insulator (b) It would melt
9 (a) water, cream, syrup, tar (b) to make it less viscous (c) wallpaper paste, custard
12 (a) 19.6 kPa (b) 16 kPa

ANSWERS TO CHAPTER 10 QUESTIONS

1 (a) particle (b) wave (c) particle
3 (a) AB, AC, BC (b) BC

4 (a) halved (b) more detailed

5 (a) Mass loss = energy loss/c^2 = 1.1 × 10^{-36} kg, which is much less than the mass of the atom

6 4.4 × 10^9 kg

ANSWERS TO CHAPTER 11 QUESTIONS

1 pressure, voltage, ions, ions, atoms, ions, ion, ion

2 (a) increased deflection (b) reverse deflection

4 (a) 0.41 per second (e) beta

5 (a) 250 million (b) 62.5 million (c) 238

6 (a) 0 (b) +2e (c) −e

7 (a) −e (b) 0 (c) +e

8 (a) uud (b) udd

9 d changes to u

10 uuu

ANSWERS TO CHAPTER 12 QUESTIONS

1 38 p, 58 n

2 (b) 7.4 MeV

3 (b) 10^{24}

6 approx 1500 kg

ANSWERS TO CHAPTER 13 QUESTIONS

6 (b) Red shift is proportional to distance
   (c) The Universe is expanding

7 (a) 400 million light years (b) 80 000 km/s.

9 Big Bang, Quarks and antiquarks, Neutrons and protons, Atoms, Galaxies

10 (a)(i) about a minute (ii) 100 000 years old

Index

light-emitting diodes (LEDs), *111*
light waves, *140–1*
lighting circuits, *121*
line spectra, *183–4*
liquids, *169*
loads, *50–2*
loft insulation, *71*
low-temperature physics, *247–8*
lunar gravity, *231*

magnetic field lines, *114*
mains electricity, *120–1*
Manhattan Project, *211*
mass, *9, 35, 41–2, 43*
mass defect, *214*
materials
    atoms, *155–8, 182–4*
    bonds, *155, 159–63*
    fluids, *169–71*
    general, *153*
    molecules, *154–6, 169–71, 175–7*
    new, *245–6*
    pressure, *172–7*
    solids, *164–8*
measures, *6–8*
megawatts, *92*
melting point, *74–5*
Mercury, *231, 232–3*
mercury thermometers, *66*
metallic bonds, *163, 165*
metals, *165–6*
metres, *7*
microwaves, *149*
mirror images, *128–30*
molecular bonds, *163*
molecules, *154–6, 169–71, 175–7*
momentum, *36, 38, 43*

motion
    acceleration, *26–31*
    laws of, *35–43*
    measuring, *17–21*
    straight line, *21–4*
    velocity, *24–6*
motor effect, *114*
motors, *114–16*
mouns, *188*

neutrons, *156–7, 158, 207*
Newton, Isaac, *34, 40–1*
Newton's laws of motion, *35–43*
Newton's theory of gravity, *229–31*
Newton's theory of light, *134–5*
Northern Lights, *196*
nuclear energy, *54*
nuclear fission, *211, 212–16*
nuclear fusion, *221–3*
nuclear physics, *211*
nuclear reactors, *91–2, 211–12, 217–20*
nucleus, *156, 158, 198*

ohms, *109*
oil, *91*
optical fibres, *245*
output power, *52*

parallel circuits, *108–9*
particles, *205–8*
pascals, *172*
perihelion advance, *232*
Periodic Table, *155, 160*
philosophy of science, *16*
photoelectric effect, *146–7*
photoelectricity, *146–9, 180*

Credits

Front cover: © Ingram Publishing Limited

Back cover: © Jakub Semeniuk/iStockphoto.com, © Royalty-Free/
Corbis, © agencyby/iStockphoto.com, © Andy Cook/
iStockphoto.com, © Christopher Ewing/iStockphoto.com,
© zebicho – Fotolia.com, © Geoffrey Holman/iStockphoto.com,
© Photodisc/Getty Images, © James C. Pruitt/iStockphoto.com,
© Mohamed Saber – Fotolia.com